Reliability Concepts in Engineering Manufacture

Reliability Concepts in Engineering Manufacture

R H W BROOK BSc, PhD, CEng

LONDON
BUTTERWORTHS

THE BUTTERWORTH GROUP

ENGLAND
Butterworth & Co. (Publishers) Ltd
London: 88 Kingsway, WC2B 6AB

AUSTRALIA
Butterworth & Co. (Australia) Ltd
Sydney: 586 Pacific Highway Chatswood, NSW 2067
Melbourne: 343 Little Collins Street, 3000
Brisbane: 240 Queen Street, 4000

CANADA
Butterworth & Co. (Canada) Ltd
Toronto: 14 Curity Avenue, 374

NEW ZEALAND
Butterworth & Co. (New Zealand) Ltd
Wellington: 26–28 Waring Taylor Street, 1

SOUTH AFRICA
Butterworth & Co. (South Africa) (Pty) Ltd
Durban: 152–154 Gale Street

First published in 1972

ISBN 0 408 70354 7 Standard
0 408 70355 5 Limp

Filmset by Keyspools Ltd, Golborne, Lancashire
Printed in England by C. Tinling & Co. Ltd
Prescot and London

Preface

Reliability is a very saleable commodity in any industry, and a customer is entitled to the level of reliability for which he or she is prepared to pay. In this book we will examine ways of obtaining greater product reliability at a given cost, sacrificing a number of sacred cows as we go along. Reliability engineering is for ordinary engineers, be they Production, Mechanical or Design. Only on rare occasions is advanced mathematical expertise required. Exercises are included to prove that either the mathematics is simple, or that simplified charts are available to reach an answer quickly.

Concepts of reliability stemmed largely from the electronics and weapon industries, but were then taken up by mechanical engineers working in the fields of nuclear energy and aerospace. In this book we see that these concepts can be taken into many other situations in a wide variety of industries. The probabilistic approaches used in electronics design cannot be translated directly into mechanical terms which would enable absolute values of reliability to be predicted. Consequently some engineers have failed to see the possibilities. However, as is shown in this book, a probabilistic approach can help decide whether one design will be more reliable than another, or whether one method of manufacture will produce a more reliable product than another. The constant theme in every chapter is that if you want to make decisions leading to a more reliable product then you must collect facts. If you do not think it is worth the effort of collecting a few facts, then may be your competitors will. Now read on . . .

R. H. W. Brook

ACKNOWLEDGEMENTS

For permission to publish data relating to their Fans, the Author wishes to thank the directors of Woods of Colchester Limited.

Contents

Chapter 1 What is new about Reliability? 1

1.1 Putting Quality and Reliability into context

1.2 Have we not always tried to achieve high Reliability?

1.3 When can the 'numbers game' be usefully employed?

1.4 Almost everything is a distributed phenomenon

Chapter 2 Material properties and distributed phenomena 15

2.1 Most material properties are distributed phenomena

2.2 The effect of distributed material properties on Reliability

2.3 The effect on Reliability of loading variability

2.4 The importance of the specification

2.5 If only I had more material data

2.6 Do not waste time around the fatigue limit

2.7 Are material properties important?

Chapter 3 A Reliable design 31

3.1 Specification

3.2 Why change a known Reliable design?

3.3 Making information available to the designer

3.4 Swallow your pride and start walking

3.5 Reliability and Maintainability checklists

3.6 We've had a failure, so we had better design a new one

Chapter 4 Failure analysis and mathematical models 43
4.1 The case for failure analysis
4.2 Your chance to try failure analysis
4.3 Reliability logic and mathematical models
4.4 Reliability logic when you have no data

Chapter 5 The development stage and test programme design 58
5.1 What is development all about?
5.2 Designing test programmes
5.3 Why does unit No. 3 perform better than unit No. 6?
5.4 Take the decision before the test
5.5 Minor development problems become major service problems
5.6 Looking for a difference
5.7 Will the test result influence the decision?
5.8 Testing with artificially damaged components

Chapter 6 What did you expect and what did you find? 76
6.1 Heads or Tails?
6.2 A simple mathematical approach to confidence

Chapter 7 Maintainability and Reliability interaction 87
7.1 Conflicting ideas
7.2 On condition maintenance
7.3 Quantitative trade-offs of *R* and *M*
7.4 Theoretical relationships between *R* and *M*
7.5 The fleet operators' view of Maintenance

Chapter 8 Analysis of service data 98

8.1 Feedback and feedforward from service experience

8.2 Problem indication

8.3 In service Reliability assessment

8.4 Detailed analysis

8.5 Weibull distribution

8.6 Service data in perspective

Chapter 9 Modelling and simulation 114

9.1 Simulation

9.2 When and by whom should models be used?

9.3 Do you believe the answers?

9.4 Putting the principles into practice

9.5 What of the future?

Index 129

1

What is new about Reliability?

1.1 PUTTING QUALITY AND RELIABILITY INTO CONTEXT

Quality and reliability are often taken to be almost the same thing, so we need to first of all understand what we mean by reliability. The Ancient Britons knew all about it with their sophisticated bridge building techniques (Figure 1.1).

A quality problem arises when, although the basic design is satisfactory, trouble is experienced in service. If the Ancient Briton used a dead tree, it would produce a quality failure in time, for whilst a 0·25 m diameter tree was satisfactory to the design engineer, he had not anticipated that the bridge manufacturer would use such substandard material as a dead tree (Figure 1.2).

On the other hand, if the contract specification had asked for a 0·05 m diameter tree and a failure occurred when using a sound tree —we have a reliability problem. The designer produced a design which failed even when using satisfactory material and manufactured in a sensible manner. Clearly the Ancient Briton Design Management were at fault, and Production could not be blamed (Figure 1.3).

This unreliability need not have happened, using the same material, but a different, and more reliable design (Figure 1.4).

There are many examples where the difference between quality and reliability is not so clearly defined. For example, the author was fatigue testing some steel specimens and some failures were occurring in the thread and others in the gauge length. Polishing the threads resulted in failures always occurring in the gauge length, so either the specimen design was acceptable, and the technique of die cutting the threads was not, or vice versa. This was a case of a reliable design with one quality standard and an unreliable design with a second quality standard.

So much for the difference in principle between quality and

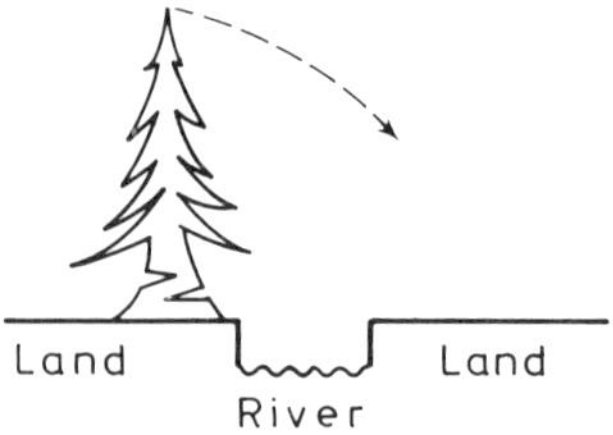

Figure 1.1 Sophisticated bridge building technique of the Ancient Britons

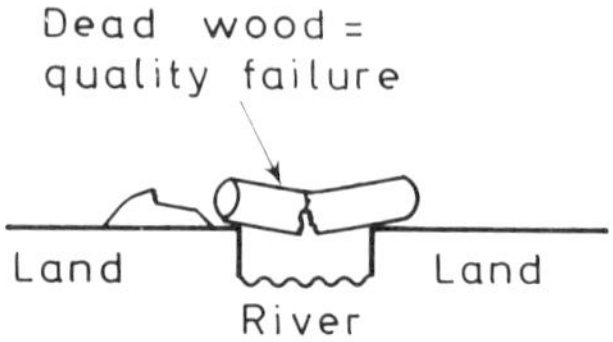

Figure 1.2 Quality failure of primitive bridge

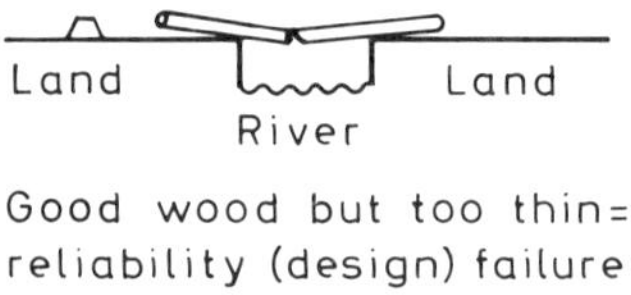

Figure 1.3 Reliability (Design) failure of primitive bridge

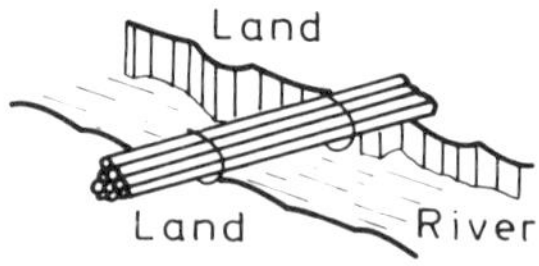

Figure 1.4 Reliable design of primitive bridge using same materials as for Unreliable bridge

reliability—now let us look at the relative importance to the customer. For a sophisticated engineering product such as a gas turbine engine, figures based on 45×10^6 hours of in-service operation, show only 9% of the unscheduled engine removals applicable to quality defects.

Consider now a much less complex and less carefully controlled manufacturing industry—house building. Two of the houses that the author has lived in were brand new and developed the following faults (Table 1.1).

Table 1.1 HOUSING DEFECTS

House 1	*Quality Problem*	*Design or Reliability problem*
large plaster cracks	poor plastering	
cooker connection not fitted	not built to specification	
paint peeled off panelling	not primed (not to specification)	
wood rot within 3 years	–	damp course designed too low
windows would not open	clearances not maintained	
cistern overflowed	pre-nylon washer standard, inlet pipe should have been throttled	

House 2	*Quality Problem*	*Design or Reliability problem*
wastepipe disconnected from sink	pipe too short	
mould on wallpaper	masonry damp from wet weather during building	
2 or 3 gate catches failed		corrosion failures – material incorrectly chosen by designer
sewer overflowed	inadequate fall of pipe	
	8	2

Here the situation is reversed and the customer would benefit from tighter quality control during manufacture (on the building site) rather than by employing more or better designers (architects).

Table 1.2 CAR DEFECTS

Problem	*No. of Occurences*	*Quality*	*Reliability*
number plate corrosion and filler cap corrosion	1	no primer	
rubber mats wore through in 3 months	1	poor material quality	
constant velocity joints failure	2	–	Life limiting feature
Sub-frames collapsed	2	–	inadequate material thickness to cope with corrosion
fuel pump failed	1	points made of inadequate material	or ?
synchromesh failed	1	–	common to all cars of this standard—cured by subsequent modifications
water pump failed	4 (1—old age)	2 failures occured at very low mileage	
steering rack failed	3	2 failures occured at very low mileage	
Throttle cable stuck	1	–	Cured by nylon liner modification
		4	4
		+1 'don't know'	

In between these two extremes of technical complexity lies the motor car. Here, based on a sample of 3 new cars of the same make and model, bought over a period of 7 years, are the faults that developed (Table 1.2).

These last two samples are too small to draw very firm conclusions, but by adding in the experience of the reader it is possible that the following conclusions will be confirmed:

Very technical product	– 80—95%	of trouble is reliability
Medium ,, ,,	– 50%	,, ,, ,, ,,
Low ,, ,,	– 5—25%	,, ,, ,, ,,

There is no substitute for facts, so to determine the size of the reliability problem in your particular industry or firm:

(1) collect the available facts on what trouble the customer experiences
(2) analyse these into quality and reliability
(3) repeat the exercise in terms of cash

1.2 HAVE WE NOT ALWAYS TRIED TO ACHIEVE HIGH RELIABILITY?

Reliability means different things to different people and several definitions have been produced, each tailored to a particular type of industry or operation. At the one end of the scale there is the 'one-off' situation, typified in history by Noah. He was given only one chance to produce a reliable vehicle for a specific task, namely an ark to ride out a storm lasting 150 days. Fortunately for Noah and his family, the design specification was given correctly, (Figure 1.5)—

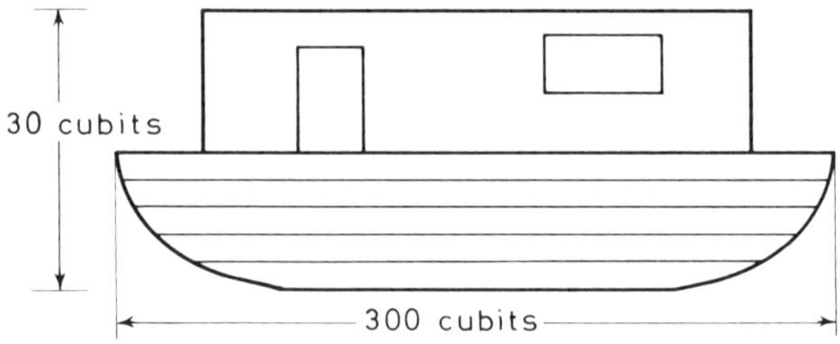

Figure 1.5 One Reliable vehicle—simple design with correct specification

'Make thee an ark of gopher wood; rooms shalt thou make in the ark, and shalt pitch it within and without with pitch[1]'.

It seems likely that the designer of the ark had made an inspired guess that 'pitch within and without' was sufficient to ensure that the ark would achieve its one-off mission. The type of mission and mission time were known in advance, but there was no opportunity for testing the design, materials or components. The penalties for manufacturing a reliable product under these circumstances was that the ark was probably overweight, incredibly expensive, and would probably have sold only at an uneconomic price. Nevertheless, the ark was unquestionably a vehicle with a reliability of 100%. The design was reliable due to simple construction, suitable material (gopher wood), minimum cutouts (one door and one window) and suitable finishing treatment (pitch).

At the other end of the scale, a manufacturer of motor cars producing a million cars of a particular model, will claim to have produced a reliable motor car if each car breaks down only once in its lifetime. In this case, a failure of each and every specimen in the batch could be regarded as a very acceptable level of reliability. One *could* regard this as a case of 100% unreliability, but no motor manufacturer would agree with that assessment. It is doubtful if many proud owners of gleaming motor cars would accept such a definition either.

This is ridiculous. Two different types of vehicles—one never failing and every sample of the other failing at least once—yet both are termed reliable.

The missing factor is time. Time may be measured in days, hours, miles, journeys, etc. To make sense of all this, the number of failures must be related to time. For the two cases considered the ratio, number of failures/time, (*failure rate*) is typically as follows (Table 1.3).

Table 1.3 FAILURE RATE

	Definition of time	*Ark failure rate (actual)*	*Car failure rate (approximate)*	*Ark reliability*	*Car reliability*
A	mile	0 per mile	0·00002 per mile	1·0	0·999 98 in any one mile
	day	0 per day	0·0005 per day	1·0	0·999 5 in any one day
	journey	0 per journey	0.0002 per journey	1·0	0·999 8 in any one journey
B	owner	0 per owner	0·3 per owner	1·0	0·7 for any one owner
	useful life	0 per useful life	1·0 per useful life	1·0	0·0 for any one useful life

Now it begins to make sense. In terms of 'per mile', 'per day', or 'per journey' the reliability of an ark or a car do not look too dissimilar, and it is only in terms of 'per owner' or 'per useful life' that the ark shows as being significantly the more reliable of the two vehicles. The differences between A and B in Table 1.3 give an indication of when the *numbers game* can be usefully employed. If the number of units being manufactured is fairly small, and the exposure or time for which those units will be used is small, then failure rate concepts may not be very useful. If the number of units is large, or the time for which the units are to be used is large, then failure rate concepts can be useful. It is instructive to think of failure rates for power stations (small number of units operating for a long time) just as much as for matches in a box (large number of units with a short operating time). This type of thinking leads to a definition of reliability as follows:

The reliability of a piece of equipment is the probability that the equipment will operate within the design specified limits for a particular period of time when used in the manner intended.

1.3 WHEN CAN THE NUMBERS GAME BE USEFULLY EMPLOYED?

Once it is accepted that a device is not going to be 100% reliable for the whole of its useful life, then probability concepts can be introduced. Returning to the car example above, if the average journey is ten miles, and on the average a roadside breakdown occurs every 50 000 miles, then there is on average a probability that 4999 journeys out of 5000 will be completed without a breakdown. It would indeed be a cautious motorist who would not accept these odds and would not set out on a journey. On the other hand, the one in 5000 chance is sufficient to induce motorists to join the AA or the RAC just in case it is their turn next. The motorist accepts the very small risk that he may break down, but often does not accept the risk that it will never happen to him and insures against this eventuality.

The numbers game approach can be adopted in any design, manufacturing or operational situation where there is some risk of failure, but if it can be shown that this risk is reasonably small, then the reliability is acceptable. Where the consequences are not acceptable, the numbers game approach can indicate the effect of carrying spares. The chances of a ship sinking are very small, but the chances of survival if that happened are negligible unless lifeboats are carried. The chances of a ship sinking and its lifeboats sinking after a successful launch are remote, and the reliability of the system, *ship + lifeboats*, is satisfactory. The car analogy is the spare wheel. The chances of a tyre going flat are perhaps 1 per motorist per year. The chances that one of the remaining tyres (including the spare) may go flat whilst the first is being repaired is acceptable, say 1 in 365 if it takes one day to repair the puncture.

The numbers game approach in the reliability of a design, manufacturing operation or in-service situation is often adopted instinctively by engineers, business men, salesmen and operators. They feel that 'it will be all right on the day' or take 'a calculated risk'.

Unfortunately 'a calculated risk' is often a euphemism for a blind guess, and this is inexcusable in most cases. Some basic data is needed before a numbers game approach can be adopted, and a designer, manager or executive can only quantify the decision if he has relevant information.

Let us consider the following design problem:
A flexible shaft inside a flexible casing transmits the drive from a motor to two screwjacks (Figure 1.6).

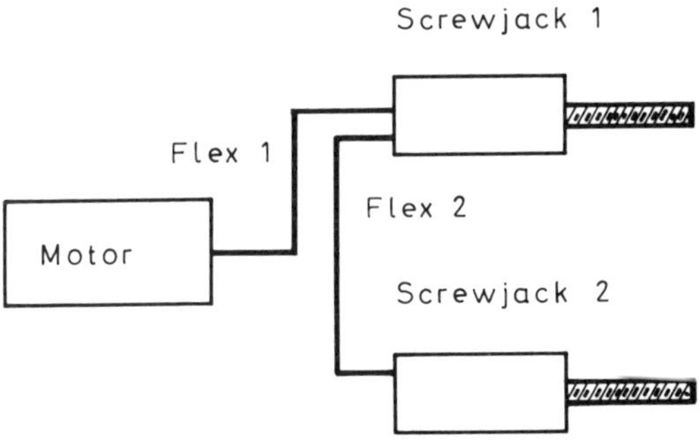

Figure 1.6 Flexible shaft transmitting drive to two screwjacks

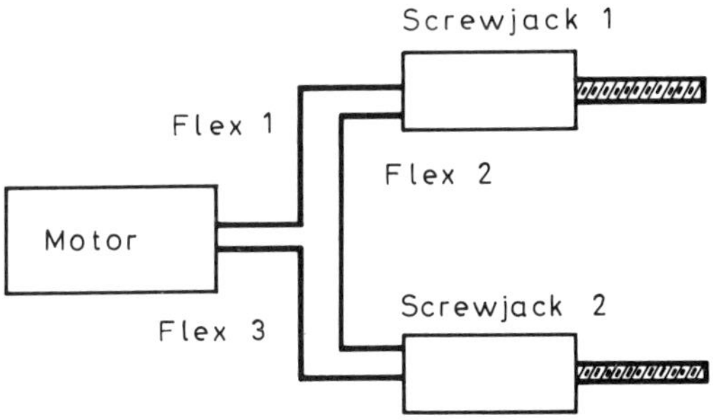

Figure 1.7 Flexible shaft transmitting drive to two screwjacks with redundant drive

If the flexible shaft *flex 2* were to break, *screwjack 2* would fail to operate. In the interests of reliability, the design could be modified by the addition of a third flexible drive, *flex* 3 (Figure 1.7).

In this way, if any one flexible shaft breaks, both screwjacks would continue to operate and the reliability of the system would be very high. This is an instinctive and not a numbers game approach. The latter would be as follows:

An investigation is made of the failure rate of flexible drives due to the drive shearing. If data shows that the failure rate is acceptably low (say f_1 per 1000 hours), then there is no need to spend money on a third shaft. If the failure rate due to shearing is not acceptably low (say F_1 per 1000 hours) an investigation of the failure rate due to the flexible shaft jamming on the outer casing is required, and suppose this is j_1. The total failure rate for the two shaft drive design will be $2F_1 + 2j_1$, and for the three shaft design will be $3j_1$ (since redundancy takes care of the shear type failures). The three shaft design is only more reliable if:

$$3j_1 \leqslant 2F_1 + 2j_1$$

In practice $3j_1$ may well be more than $2F_1 + 2j_1$, and so the 'reliable redundant design' is more expensive, fails more often and is less reliable than the 'simple less reliable design'.

As a second example of the numbers game approach some stainless steel bolts were failing in service due to fatigue cracks starting at the bottom of the threads. Alternative approaches to solving this problem are:

(1) fitting bigger bolts to reduce the load on each bolt
(2) improving the manufacturing techniques by adding a thread polishing operation

The first option implies redesign, introduction of a modification which may not allow rework (modification of the existing hardware) with all the associated expense that this implies. The chances of a successful cure are very high. The second option implies minimum change to existing hardware and design, but there is less chance of obtaining a complete cure.

A numbers game approach can be used to enable the correct decision to be made, provided data is available for assisting the decision. The data is likely to be of the following form:

1000 units have been in operation for an average time of 2 years, and 10 have failed in the manner described above.

In other words, only 1% of the units have failed so far, so only a modest improvement is needed to improve the reliability to a satisfactory level and to stop further failures occurring in service. Some quantitative information is required about the affects of thread polishing, and a fatigue test of three bolts made by the old process (process 1) and three bolts made by the new improved process (process 2) including the thread polishing operation, would provide this necessary information. Typical results would be:

Bolts made by process 1. 10 000, 15 000, 20 000 cycles to failure
process 2. 50 000, 75 000, 100 000 cycles to failure

This shows on average that the life of bolts made by process 2 is improved by a factor of 5 relative to three bolts made by process 1. There is therefore a good chance that bolts of process 2 will fail at the rate of 1% in 10 years instead of 1% in 2 years, and very likely this will be an adequate improvement.

This looks to be a trivial example at first sight, but it does emphasise that only a modest amount of data may be needed to make the right decision. Absolute values of reliability or absolute estimates of component lives may be impossible to test for or to estimate, but comparative data is usually all that is needed.

The reader is now beginning to see why he does not often need to be a statistical wizard to adopt a numbers game approach. It is usually rare that enough data is available to satisfy the pure statistician and it is usual to have to settle for a combination of quantitative facts plus engineering intuition. Moreover, where engineering judgment gives a completely different answer to the statistical data, then the quantitative data should be reexamined. It well may be that the data being provided is unintentionally one sided, and will only give the *bad* data, and not reveal the *good* experience that has been accumulating at the same time as the bad.

'I lose £1000 every year due to careless handling—I despair of ever teaching those chaps the importance of taking care with part manufactured goods' is the bad side of the story put across by the shop foreman. The correct reply to this is often, 'Yes, but there is £250 000 of part manufactured components that do not get damaged by careless handling', or in reliability terms, the handling in that shop is no less than 249/250, or 99·6%. This second way of looking at the facts presents the balanced story and allows Management to make the correct decision, which in this case would be to leave well alone as the operation is going very well indeed. This theme will be further developed in Chapter Three, where we examine the typical case known to us all, namely the enthusiastic production engineer or designer who comes rushing into the office looking overwrought: 'Process X is no good, we've had a scrap component made by it only yesterday'.

The designer's equivalent is:
'We've had a failure of that component, I had better redesign it straight away'.

The production engineer has inbuilt motivation to change a process for a better one, and the designer has inbuilt motivation to change any design for a better one.

The true reliability of the process or design can only be properly assessed by looking at the failed and the non failed components. A change is not by definition necessarily for the better.

1.4 ALMOST EVERYTHING IS A DISTRIBUTED PHENOMENON

This is well known to the Quality Engineer, but for the rest of us let us spend a minute or two thinking. The simplest example is a case where there are only two possibilities, as happens when you toss a coin. If a coin is tossed ten times, most people would be very surprised if they did not get a single tail, and fairly surprised if they did not get two tails. The following tabulation is the results of ten people each tossing a coin ten times on ten different occasions.

No. of tails in ten tosses	*No. of occurrences in* 100 *trials*
10	0
9	1
8	3
7	14
6	32
5	24
4	17
3	6
2	2
1	1
0	0

In this particular experiment on 1% of the occasions only did one tail appear during ten successive tosses. If that had been an experiment in a machine shop or an experimental shop, how many readers would have doubted, that as a result of no less than ten tests, that the chances were that the outcome was nearly always heads. In practice many Engineers may have found a very plausible reason for discounting the odd result when tails appeared and reached the conclusion that heads was always going to be the outcome.

Fortunately this 1 tail in 10 tosses would happen only once in every 100 tests and once in every 100 tests there would be only 1 head, on the average, and 98 times out of a 100 the results would not have been so misleading. However, it is worth noting that only once in four test series does the answer appear completely 'correct', and the same number of heads appear as tails. So, notwithstanding the fact that *on average* the coin lands tails as often as heads, three times out of four the answer from these tests was different from this.

Another way of looking at this is to realise that if a decision has to be made between two alternatives, then 50% of the time the correct answer will be given by tossing a coin. In other words, if your Manager is not correct more than half the time, fire him and toss a coin instead. This will cost you nothing and give you better decisions.

Hopefully, the Managers working for you are correct more than 50% of the time, and are therefore more reliable than a coin, but is it not worth checking their record? There is more coin tossing in Chapter 6.

Not only are the results of tossing coins, and the accuracy of decisions distributed phenomena, but almost everything else as well. Material properties are good examples. It is very common knowledge that the fatigue properties of materials are a distributed phenomena, but it is less often realised that the common static properties such as Ultimate Tensile Strength and Hardness are also distributed phenomena. Examples of the three distributions of these properties for one particular steel are shown in Figures 1.8,[2] 1.9, and 1.10. These distributions of properties refer to specimens from a single cast of steel.

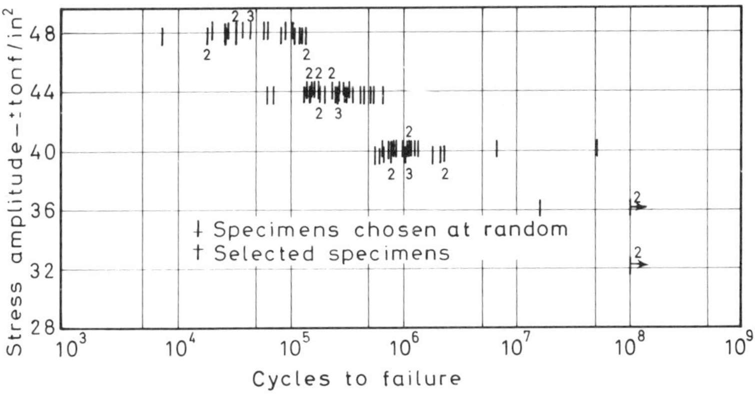

Figure 1.8 Fatigue properties for plain specimens of stainless steel Rex 535

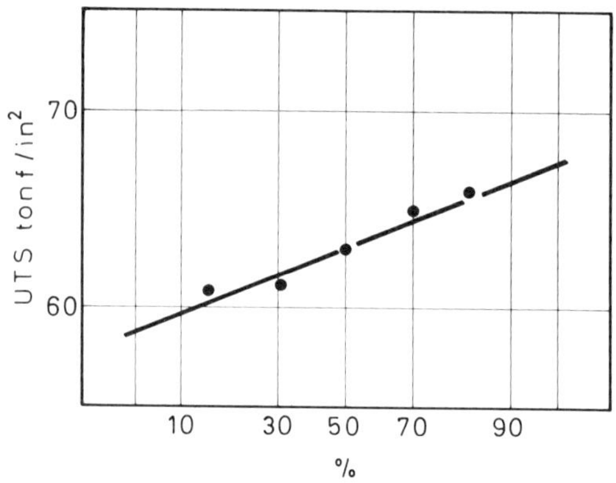

Figure 1.9 Distribution of Ultimate Tensile Strength for stainless steel Rex 535

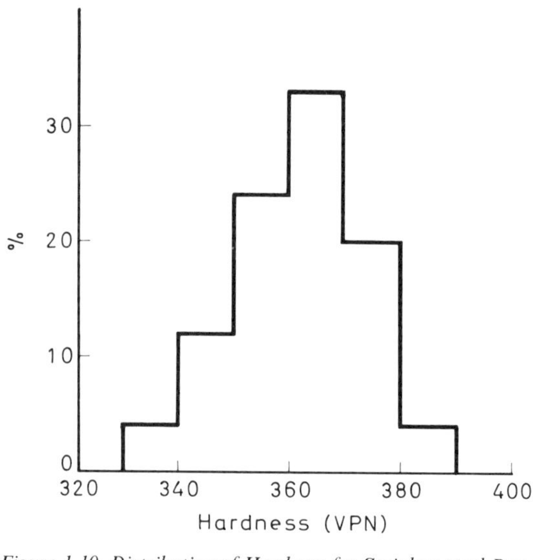

Figure 1.10 Distribution of Hardness for Stainless steel Rex 535

The motor car demonstrates to us the distribution of performances that occur with machines due to the combined effects of the distributed tolerances on materials, on the dimensions of the manufactured components, on the quality of the welds, on the tightness of the belts and so on. Examples from the author's motoring experience are:

Component	*Lives to failure (measured in miles run)*
steering rack	4 000, 5 000, 11 000, 75 000+
water pump	3 000, 7 000, 10 000, 65 000

These are components to the same design and typically the ratio of best life to worst life is 20:1. Fortunately there are not many other components that fall within the author's experience of motor cars that have provided so many failures examples. The other failures are more random, but only because the sample size is small. If other owners of similar cars added their experience, then a pattern would emerge.

The dimensions of a component produced either by manually controlled or automatic machines will be distributed in one way or another. An example of the distribution of the dimensions of a diameter turned on a hand operated lathe using several different operators and subsequently polished is shown in Figure 1.11. From

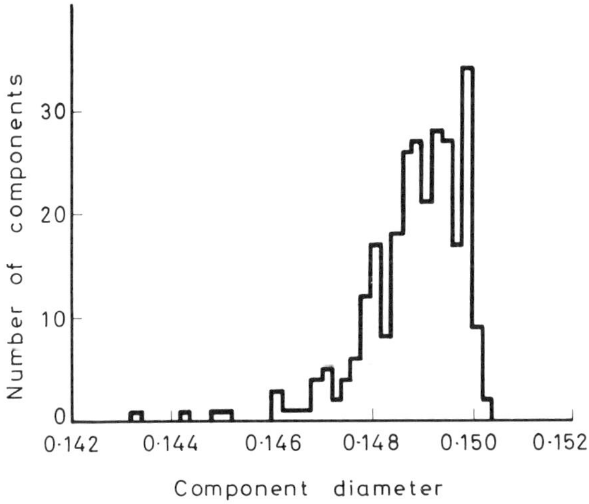

Figure 1.11 Distribution of diameters of component following hand turning plus polishing; turned dimension to be 0·150±0·001 but polishing continued until specified surface finish achieved

this example it can be seen that the distribution of diameters is far from normal and the polishing operation has given the distribution of diameters a long tail at the thin end. It also suggests the preference for the operators to check the dimension just before the nominal 0·150 inches and then to wind in a bit further to save a second check.

In many cases it is quite sufficient to collect the available facts and analyse them in a simple way. In trying to decide whether or not to believe a series of test results or a collection of production data, relating the apparent answers to simple concepts of coin tossing, drawing cards from a pack, or looking at a normal distribution table or a Poisson chart (Chartwell graph sheet No. 5591A) may be all that is required. On other occasions, perhaps only once in fifty times that a decision is required it will be cheaper to call in a statistician. There will be occasions when he can actually be useful, but on many other occasions common sense plus experience in the relevant engineering field will provide all the answers that are required. So often the question is 'do I do it this way or that way'? and absolute quantitative answers are not required. In trying to answer this question, there is no substitute for a few relevant facts and an appreciation that almost everything is a distributed phenomenon.

2

Material properties and distributed phenomena

2.1 MOST PROPERTIES ARE DISTRIBUTED PHENOMENA

Several examples of distributed material properties have already been given in Chapter One, but to reach a deeper understanding of the unreliability seen in service or on test, this aspect of material behaviour must be studied further. The literature of fatigue is full of examples of the scatter to be found in the cyclic properties of material, and naturally a number of workers such as Freudenthal[3] and Weibull[4] have studied carefully the statistical nature of fatigue.

The scattered nature of fatigue properties is demonstrated forcibly by Nixon[5] who has published a combination of original and republished work (e.g. that of Cicci[6]) to show that the ratio of maximum to minimum life under given conditions of repeated loading can easily be 10:1, 100:1, or even 1000:1. That is to say for the same material, maybe even from the same cast, under carefully controlled loading conditions, the best specimen or component tested may last for ten or a hundred times longer than the worst one tested. Nixon shows that this can be equally true for specimens manufactured from actual components and for specimens made from a single plate (Figures 2.1 and 2.2). No doubt these are extreme cases chosen to make a point, but the general and typical case will usually show a scatter in life such that the ratio of longest to shortest life for constant loading conditions is about ten. Returning to the scatter of static properties in a typical case rather than for the single cast of steel case shown in Figure 1.10, the distribution of the Ultimate Tensile Strengths of test pieces taken from real components are shown in Figure 2.3. The steel is the same Rex 535 stainless steel mentioned in Chapter One, but heat treated to give a different mean Ultimate Tensile Strength. The data is based on a 20% sample of forgings over a period of about eight years. The quality control

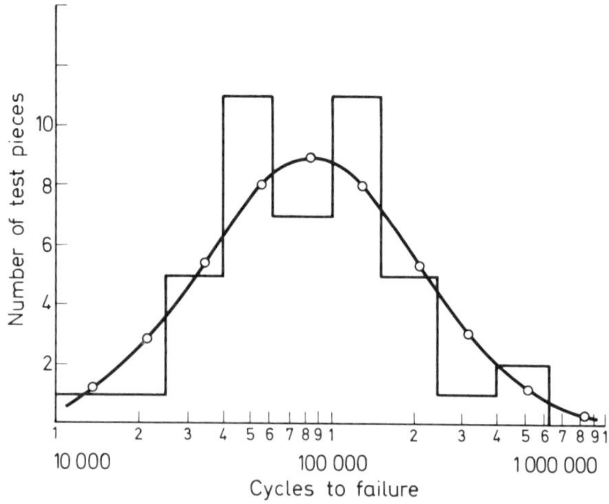

Figure 2.1 Distribution of Fatigue properties—results of tests on specimens from actual components

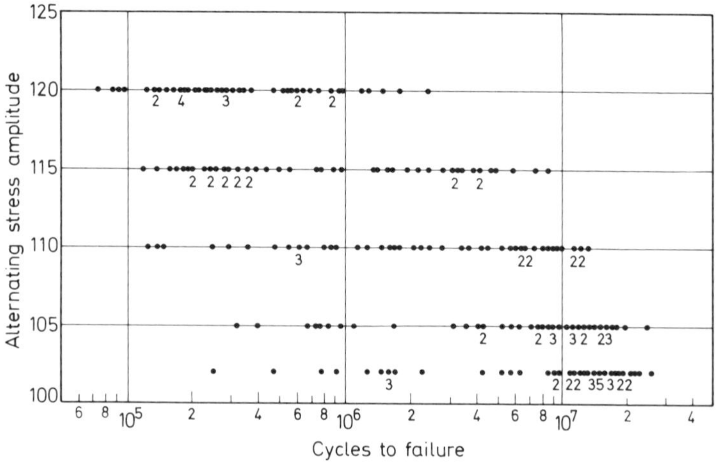

Figure 2.2 Distribution of Fatigue properties—results of rotating bending tests on a maraging steel. All specimens from a single plate, air melted, vacuum steam degassed

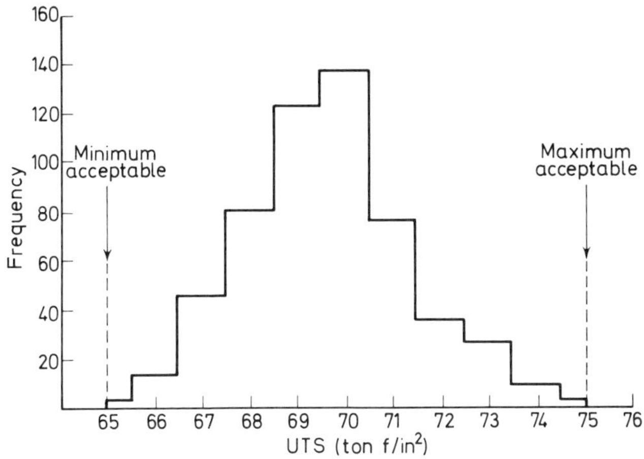

Figure 2.3 The distribution of Ultimate Tensile Strength of stainless steel (Rex 535) specimens, cut from real components, shown in histogram form

would appear to be very good for no results fall outside the limits, and the distribution looks reasonably Normal. In view of the long period of time over which this component was being manufactured it is instructive to look at a scatter plot of Ultimate Tensile Strength against Calender Time (Figure 2.4). Without getting mathematical about it, there seem to be two populations, one composed of the first production batch of twenty, the second being the rest. If this apparent fact needs to be demonstrated mathematically, then distributions of strength of the first twenty and also of the total population over the eight year period are plotted on Normal probability paper showing two quite distinct populations (Figure 2.5). Random sampling over the whole period would produce a reasonable description of the total population, but to assume that a non random sample such as the first twenty produced will describe the total population is invalid. An appreciation that material properties are always distributed phenomena would avoid invalid conclusions being drawn.

For an engineer working with very conservative safety factors or reserve factors, the fact that an Ultimate Tensile Strength may lie between 65 and 75 tonf/in^2 may be purely academic. However, in a competitive world it is becoming uncompetitive in many instances to use large reserve factors, and engineers are working in situations where the variability of material properties becomes an important factor in reaching correct decisions.

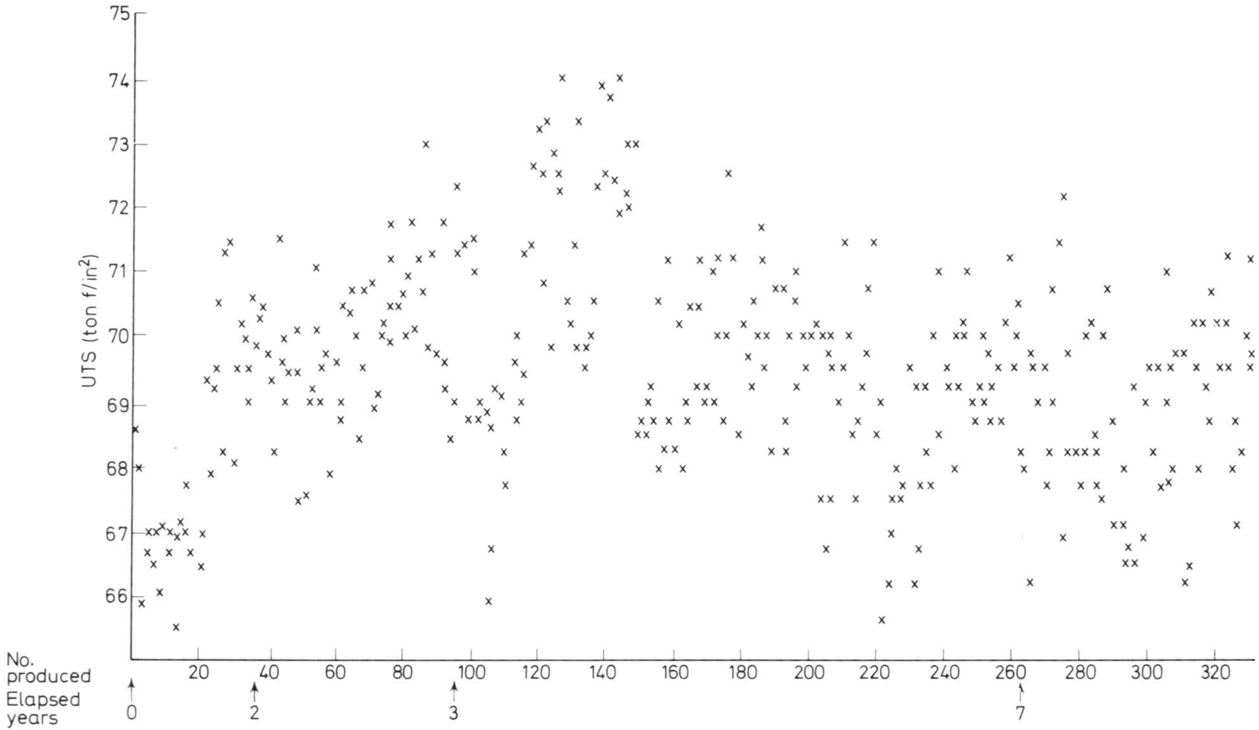

Figure 2.4 The variability with number produced and time of the Ultimate Tensile Strength of stainless steel (Rex 535) specimens, cut from real components

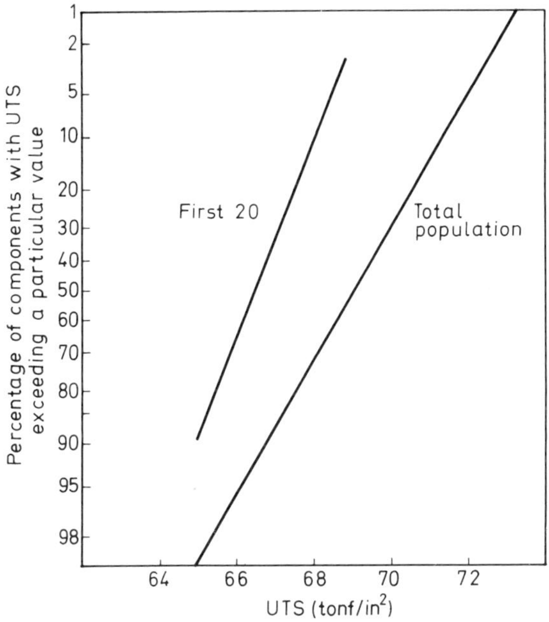

Figure 2.5 The strengths of the population of the first few components produced may be different from the strengths of the total population

2.2 THE EFFECT OF DISTRIBUTED MATERIAL PROPERTIES ON RELIABILITY

Obviously if a component is being loaded such that the applied cyclic loading is in the vicinity of the fatigue limit of the material, then the scatter in life to failure can become infinite. This is rather a special case representing the upper loading limit for a design intended never to fail (Figure 2.6). In practice a vibrating load may well be superimposed on a steady load in which case the fatigue limit may be very small. This reduction in fatigue limit is normally calculated using the modified Goodman diagram[7], although there is a line of thought suggested by Passey[8] (Figure 2.7), which would like to further modify this hypothesis by relating predicted behaviour more closely to the relevant measured material properties. Before you accuse me of forgetting that this chapter is about the distribution of material properties, Brook[9] shows that the scatter experienced in practice will in certain cases be even greater than discussed above because the fatigue limit calculated from Goodman or Passey (or anybody)

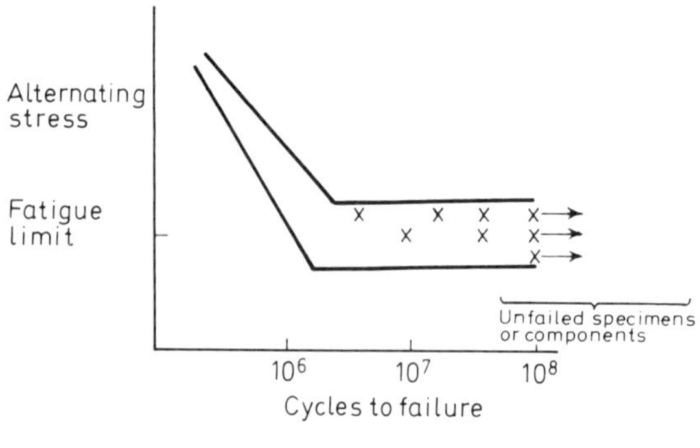

Figure 2.6 The scatter of fatigue lives near to the fatigue limit can be infinite

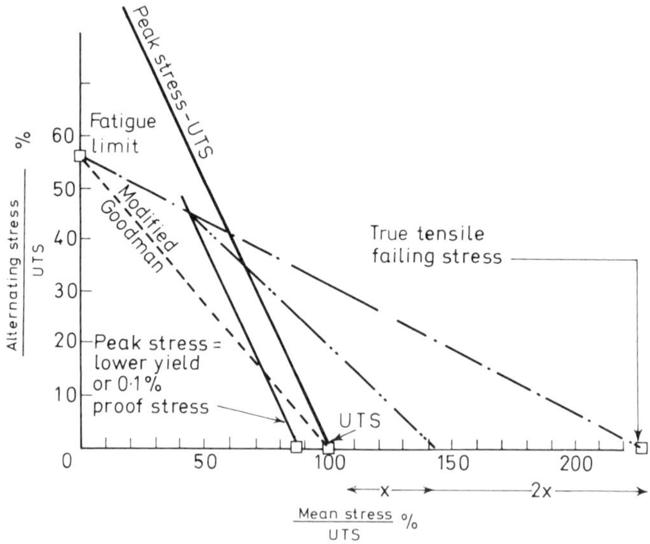

Figure 2.7 The Passey Range-Mean diagram—an attempt to avoid additional scatter by avoiding overestimation of safe alternating stresses

is not a fixed material property, but can be modified and reduced by the effect of any major cycling. Major cycling can make a material progressively more sensitive to small vibratory loads, the effect being more devastating on the 'worst' material than on the average or best material (Figure 2.8). This hypothetical curve shows how the

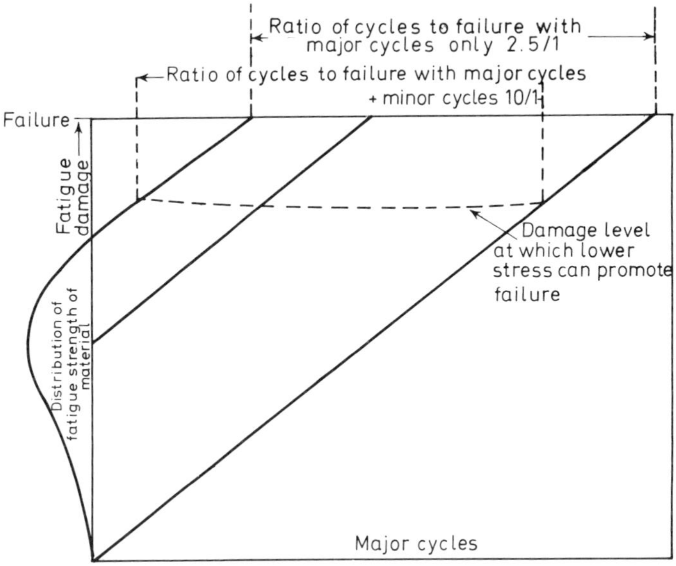

Figure 2.8 Major cycles produce a fatigue damaged state at which minor cycles can promote failure

apparent scatter in material properties may increase from say a ratio of:

$$\frac{\text{maximum life}}{\text{minimum life}} = 2{\cdot}5 \text{ to a ratio of } \frac{\text{maximum life}}{\text{minimum life}} = 10$$

when the mode of failure is fatigue failure caused by a combination of major and minor cyclic loads. Some experimental evidence supporting this claim is reproduced in Figure 2.9. There it is shown that for the particular steel investigated, minor cycles below the conventional fatigue limit can produce failure in a poor piece of material after only a few major load cycles, whereas for the average or better than average pieces of material the minor cycles have little or no effect for a very long time.

The net effect of this interaction of major cyclic loads and minor vibratory loads is that the distribution of failures experienced in practice may well be more skewed than would be expected from tests under one cyclic loading condition, and the bad specimen or components will appear even worse if subjected to multilevel cyclic loading. Distributions of failures of real components subjected to major cyclic loads with superimposed vibrations are reproduced in Figure 2.10[9] and in Figure 2.11[10]. All three components for which failure distributions are shown are gas turbine engine components, subjected to a combination of major cycles, with the maximum load usually occurring at take off or at the end of climb, and minor cycles, which may be due to aerodynamic induced vibrations or small undetectable fluctuations in the fuel control system. A similar loading

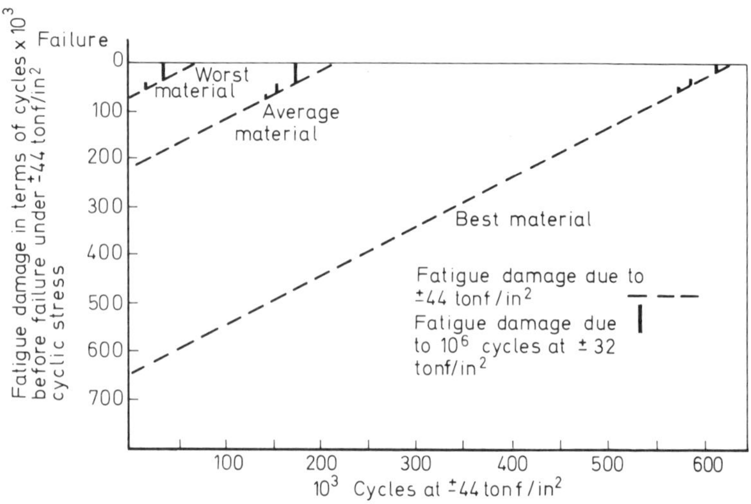

Figure 2.9 Superimposed vibration has the greatest effect on the reliability of components made from the worst material

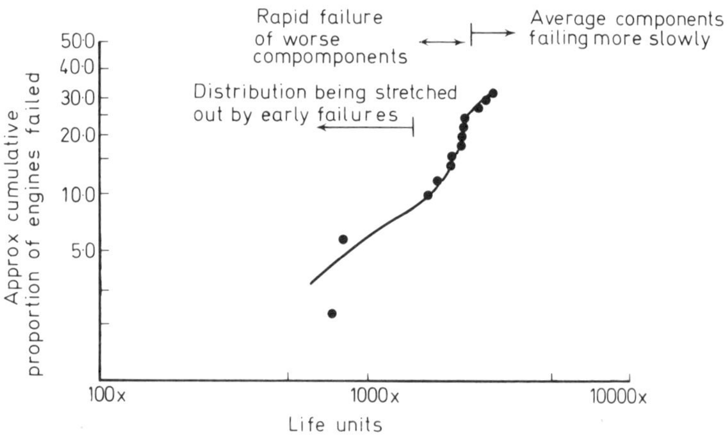

Figure 2.10 Turbine Blade Failures

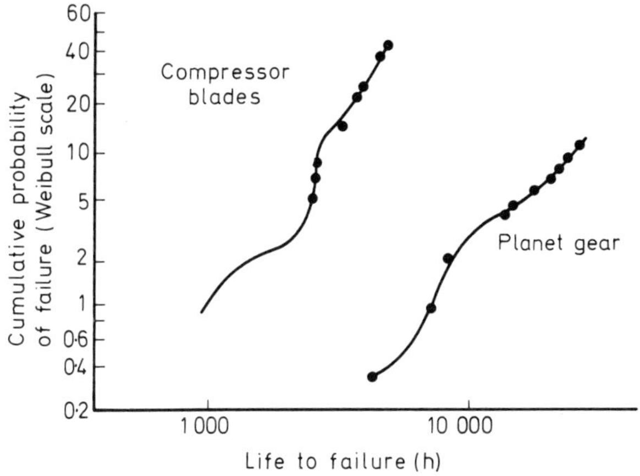

Figure 2.11 The long tailed distribution of components that fail due to a combination of high and low cycle fatigue—gas turbine compressor blades and planet gears.

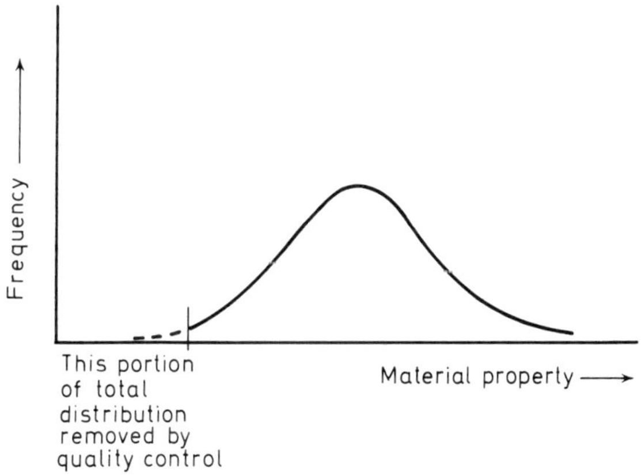

Figure 2.12 The bottom end of the distribution of a material property is often removed by good quality control

situation may apply to airframe structures, with gust loadings providing the minor cycle loading; cooling towers and bridges, with the wind forces providing the minor cycle loadings; motor car components, and indeed to many mechanical and civil engineering components or structures. Hence the justification for spending most of this chapter in thinking about failures due to cyclic loadings.

Now that a really depressing picture has been painted, why is it not all that bad in many practical cases, and why do engineers usually succeed in producing a reliable product? Fortunately, quality control and non destructive testing techniques are often extremely successful at truncating the inherent distribution of material properties. Most material properties are Normally distributed if left on their own, but no reputable Production Engineer leaves them on their own. He tries to get rid of the worst ones (Figure 2.12) in order to improve the chances of producing a reliable product. He never tries to get rid of the best ones—in fact he may even put them on one side for demonstration purposes, so it is usual to see a reasonable spread of lives or cycles to failure of the component, whatever the mode of failure.

This truncation of the bad end of the distribution is shown nicely by Bompas-Smith[10], who plots the distribution for all test pieces taken from a batch of components (Figure 2.13) and compares this

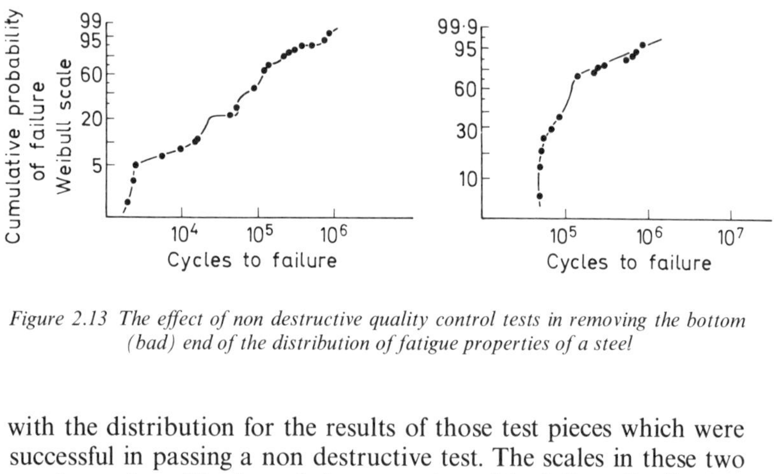

Figure 2.13 The effect of non destructive quality control tests in removing the bottom (bad) end of the distribution of fatigue properties of a steel

with the distribution for the results of those test pieces which were successful in passing a non destructive test. The scales in these two graphs are logarithmic, and the truncation of the failure distribution in this case is quite dramatic. This is precisely what every inspector in a factory is trying to achieve, although it is not usual to be able to quantify the results of his work in this way.

2.3 THE EFFECT ON RELIABILITY OF LOADING VARIABILITY

These are not the only reasons for the distributions of failure that occur in practice. The affect of applied load variability, environmental conditions, method of manufacture, the supplier of the material or of the maintenance procedures adopted can make a significant difference to the variability experienced by the user. The effect of maintenance procedures will be considered in Chapter Seven.

The cause of a failure is that in some way the load that the component or assembly is being asked to carry, whether it be cyclic or static, is too great under the prevailing environmental conditions. It is possible to show this concept in diagrammatical form as the overlapping of the distribution of possible material strengths and of the distribution of possible applied loads (Figure 2.14). However, it

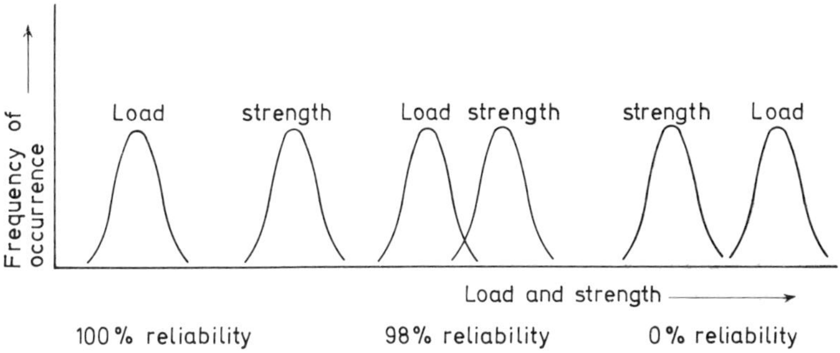

Figure 2.14 Reliability depends upon the interaction of the load distribution and the strength distribution

may be difficult to establish the distribution of applied loads accurately in many situations, although in certain special cases such as gust loading of airframe structures, considerable work has been carried out to establish the statistics of magnitude and frequency of loads that will be encountered.

There is little to be gained in overall product reliability if the material properties are known precisely but the loads applied are known only approximately. The project management have the responsibility of maintaining compatible effort between the two. Equally well it is important to understand whether the material properties that have been, or are to be established apply at the operating environmental conditions when the product is being used by the customer. So perhaps the previous few pages are just an Aunt Sally set up to be knocked down? Yes, if the envelope of loadings within which reliable operation is to take place has not been defined as a specification, and no, if it has.

2.4 THE IMPORTANCE OF THE SPECIFICATION

Whilst this chapter set off to discuss the distribution of material properties it has suddenly gone off to talking about Specifications. All very odd? If the Specification cannot clearly define the limits of the loads to be applied then all the reliability theory in the world is not going to be of much assistance. In such a case either some large safety factor must be applied to cater for the unknown loads or else it is a question of build one and see what happens. In either case the product is likely to be heavier than is really necessary. If the Specification does clearly define an envelope of possible applied loads, then an understanding of the distribution of material strengths in the manufactured state can be used to ensure that the desired level of reliability will be achieved.

This calls for co-operation and communication between Production and Design engineers to ensure that the method of production is understood by Design so that the possibility of a change in the assumed distribution of material properties can be foreseen.

This applies particularly to production techniques that can be changed more or less at the whim of the operator or shop foreman, such as welding speeds, or sheet metal forming. In this case the reliability of the product may depend upon a second type of specification, one which specifies the manufacturing processes that have been assumed at the design stage and which were used to produce the early development units. The configuration of the units to be tested must match or have only known deviations from the configuration of the units that will be delivered by production. Configuration is being used in the very widest sense to include the method of production where this may affect the distribution of material properties that will apply to the finished component or assembly.

Similarly, the specification must state what the environmental conditions will be in service. Not only will the average or mean properties of the material change with changes of temperature or corrosivity of the service environment, but the relative positions of two possible choices of material may change over. For example the effect of temperature can change the order of performance, so that alloy A is preferable at laboratory ambient temperature but alloy B is preferable at the in-service environmental temperature.

2.5 IF ONLY I HAD MORE MATERIAL DATA

If you come into contact with Design or Stress Engineers then it is likely that you will have heard the above quotation on a number of occasions. From the reliability point of view it is simpler than that. Until such time as sufficient data has been collected about the relevant material properties of a new or different material, then its introduction must be resisted. If other conditions make the adoption of a new material essential, such as achieving a competitive performance, weight, volume or cost, then the senior management must be made to appreciate that the product may be initially less reliable than the current version.

Ideally, the designer should establish a programme which shows the date by which a certain minimum amount of information must be made available in order to choose technically between alternative materials (Figure 2.15). This data must be compatible with the

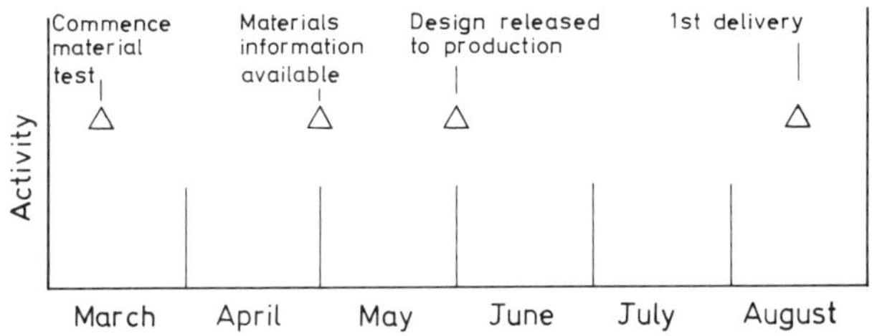

Figure 2.15 If the reliability of a piece of equipment depends upon a knowledge of material properties, then the programme must recognise this

overall production programme and delivery dates. There is a parallel situation when alternative methods of production are being considered which may have a considerable influence on material properties. This would certainly occur in considering whether to go for casting or fabrication, or in trying to decide between casting and forging.

It will never be the case that *all* the desirable material information is available. The responsibility for establishing in these situations the priorities to say which information is required first, must lie with the Designer or Production Engineer as appropriate. He is complaining from a position of weakness if he has not stated in advance of having to reach a decision what he requires to be able to make that decision. The priorities for obtaining information are, from the reliability point of view, related directly to the order in which the failure modes are ranked. For example if the component fails in service due to simple overloading, then the static properties are those of most importance. If the predominant mode of failure is vibratory fatigue starting from a severe stress concentration, and the operating temperature of the component is 300°C, then the notched fatigue limit at 300°C is the property of most interest. If the laboratory are busy looking at the plain fatigue strength at room temperature, then it is probably the fault of the designer for not setting priorities for investigation. If the production process development people are putting most of their resources into looking at ways of improving casting techniques when the designer has other constraints upon him such that his critical items are going to be fabricated, then both parties are at fault for failing to communicate. It is no good either party coming along at a late stage in the programme only to find that they are following the wrong priorities.

2.6 DO NOT WASTE TIME AROUND THE FATIGUE LIMIT

In many situations there are a limited number of alternative materials that can be used with any degree of confidence that a reliable product will be delivered to the customer. In such situations the choice can be made between competing materials if the answers of these questions are known:

(a) Which material exhibits the least scatter in the property which controls the anticipated dominant failure mode?
(b) Which material has the greatest strength (fatigue life, creep life etc., as appropriate) amongst the competing materials?

The absolute values of the properties under these circumstances may not be too important and a very modest amount of testing will sort out which material to choose.

In many mechanical situations, it is the fatigue properties which are the most important in considering the likely modes of failure. There is plenty of evidence in the literature that for a given standard of quality control, the scatter in fatigue properties for all metals is very considerable, so that question (a) above need not be answered. If the design philosophy is for no failure and no life limitation, then all stresses in the product in service must be below the fatigue limit.

However, if the choice is limited to two or three materials anyway, as is often the case, there may be no need to spend a large amount of time testing at stresses around the fatigue limit trying to establish which material has a higher limit. Provided the environmental conditions are not too dissimilar from those in service, then the material which has the best fatigue properties at high stresses and strains (and hence short total lives) will also have the highest fatigue limit.

In other words, if comparisons are all that are required it is much more economical in testing time and cost to sort out the materials at higher loads than apply in service. This is demonstrated in Figure 2.16. Material A is preferable to Material B from the fatigue failure

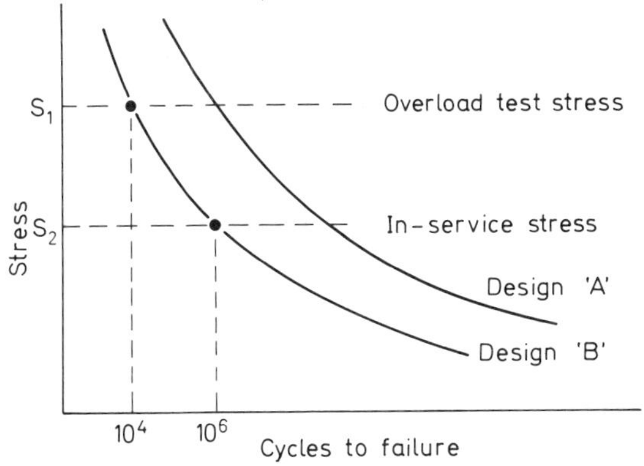

Figure 2.16 Overload cyclic tests to compare alternative designs

aspect at all loading conditions, and to sort out the preferable material by applying high loads will take only one hundreth of the time it would take by testing around the fatigue limit. Testing at S_1 rather than at S_2 will show this order of reduction in testing time. This is always assuming that the question to be answered is 'do I use material A or B?' Published data often shows an obsessive and almost unhealthy interest in properties around the fatigue limit, when as we have seen (Figure 2.6) that is where the properties are least well defined and the distribution of fatigue life is greatest.

These ideas of reducing the testing time required to make a valid decision are explored further in Chapter Five.

As for 'how many tests do I need to carry out?' in view of all that has been said to emphasise that material properties are distributed phenomena, the non statistical answer is 'quite a lot' if absolute numbers to describe the properties are required, and 'not very many' if only a comparison is required. Again, more of this later.

2.7 ARE MATERIAL PROPERTIES IMPORTANT?

A whole chapter has been devoted to talking about the variability of material properties. This is justified and reasonable because materials are one of the few things that are common to all branches of engineering. In the end, no matter what kind of specialist engineer each one of us may call himself, we are only real engineers if our endeavours ultimately produce something, whether it be a structure, a mechanism or an electronic device. In the end, an engineer has to use materials to produce something tangible, and it is as well if he realises that the property with which he is concerned will never have a unique value but will be a distributed phenomenon. A single failure may mean that the worst one in the batch has now been removed and that the rest will be satisfactory.

3

A Reliable design

3.1 SPECIFICATION

A reliable design usually arises from a balance of intuition, mature judgement, a knowledge of the material properties relevant to the in-service condition, such analysis as can be helpful (e.g. stress analysis, value analysis, reliability analysis), feedback from service experience, feed forward from the research areas and from good communication between Designer, Development Engineer, Production Engineer, Service Engineer and any other engineer involved in the project. Communication between those parties who have something to contribute is the first essential in achieving a reliable design.

The much quoted KISS system also helps to promote in-service reliability:

Keep
It
Simple
Stupid!

The Chief Designer, Reliability Engineer, Production Engineer and anyone else in on the act must always remember that the Designer's main motivation is to produce a change. If things are done exactly the same way as last time, then nobody needs a Designer! Equally well, if the components are manufactured the same way as last time, then the Production Engineer may be feeling a little bit redundant.

From the Reliability Engineer's standpoint, if it worked well enough last time then it will work well next time, and the less anyone wants to change the design concept or manufacturing method the better he likes it. Complexity and change are often the signposts on the stony road leading to unreliability and dissatisfied customers. How does one persuade the advocate of change to read these signposts? The answer must be to ensure that the specification which tells the design and manufacturing groups what they are expected to achieve includes quantitative requirements specifying what reliability must be achieved to meet the contract. Obviously this target must be realistic and related to what is achievable if it is to be credible.

Ideally, the specification should tell the designer every requirement that the piece of equipment has to meet. Some of these requirements may be contractual and must be guaranteed, others may be aims or objectives. At first sight a written specification for a piece of aerospace or other highly technical equipment may seem a formidable document.

However, a carefully prepared comprehensive specification is a boon to the designer, for then there is no doubt what the piece of equipment must do, and what must be its minimum performance, operational environment, design envelope, quality and reliability. There is an increasing tendency to specify reliability quantitatively. This can be straightforward, for example, in a complete motor car the minimum warranted reliability is often 12 months or 12 000 miles. This is simple and measurable, and at the end of 100 000 miles or so of accumulated testing it will be possible to say whether the various equipment is likely to meet the reliability targets or not.

The situation may be more complex. Consider an aircraft where a number of systems could cause a departure delay. The aircraft is sold with a certain guaranteed minimum delay rate, say 1%. By the time this guarantee has been apportioned to the various aircraft systems, the delay rate for each individual system is very small indeed and typically may be a mean-time-between-failure of $0{\cdot}25 \times 10^6$ flights, which is a failure rate of about 4 per 10^6 flights (Figure 3.1). The

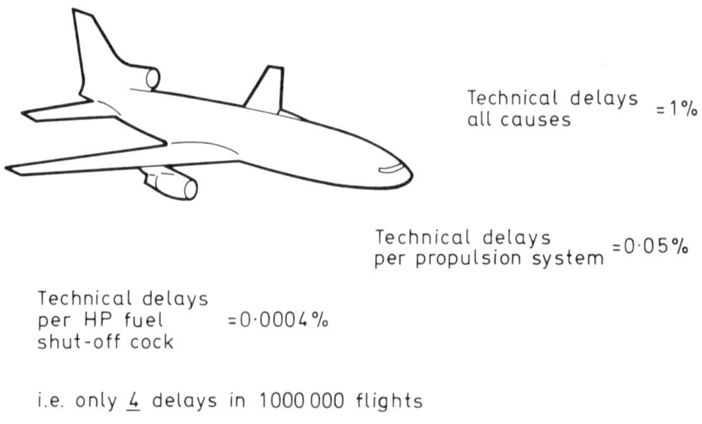

Figure 3.1 Quantitative Reliability Requirements

amount of testing to demonstrate this level of reliability would be quite prohibitive, and for any confidence at all would require 5×10^6 hours testing (more if a pure statistician is looking over your shoulder).

The fact that achievements cannot be easily demonstrated should not deter us from specifying the required level of reliability. It is usual today for houses built to approved N.H.B.R. standards to have a ten year guarantee on the fabric and no one is worried that this level of reliability will not be demonstrated in a short time. Anyone involved in contracting for the United States Government will know that in establishing major defence contracts, immense care is taken to determine the correct specification. One of the main objectives of the Project Definition Phase (PDP) is to establish firm and realistic specifications. If the specification is right then at least there is a

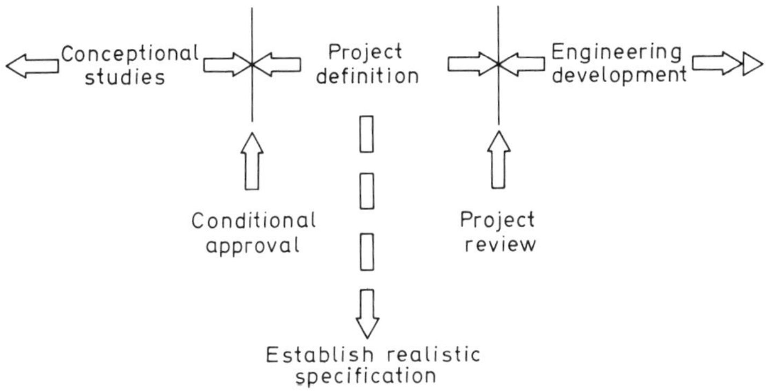

Figure 3.2 Establishing the specification

possibility of the product being right, (Figure 3.2) and the required goals being achieved.

Regarding specifications for reliability in the aero engine industry, what is tending to happen is that the customer (the airframe manufacturer) specifies the required quantitative reliability objectives for the whole propulsion system. After some negotiation as to what objectives can be accepted as realistic, the aero engine Reliability Engineers break down this overall objective into objectives for specific components or accessories. This breakdown is arrived at by studying past performance, and comparing this with the proposed design changes. In this way the designers and subcontractors know from the specification precisely what their bit has to do to enable the total propulsion system, and total aeroplane, to meet their reliability objectives.

3.2 WHY CHANGE A KNOWN RELIABLE DESIGN?

This question is best answered by considering a particular case in a particular industry. Gas turbine engines have been in commercial airline service operation since 1953. At first it would seem reasonable that the reliability of these turbines should improve continuously. The reliability improvement of an individual type of engine continues with time as service operation shows up the shortcomings and corrective action is taken to cure the problems. Maturity with a particular mark is reached in about 2 to 3 years from the first revenue flight (Figure 3.3)[12]. Why then are we prepared to go back to square one with a new engine? The answer is purely financial, and if we consider the earning capacity of an engine in terms of:

$$\frac{\text{(revenue per hour per engine)} - \text{(fuel cost per hour per engine at cruise)}}{\text{new engine cost}}$$

the situation is shown in Table 3.1.

Table 3.1 ENGINE CASTINGS

engine	*aircraft*	*entered service*	*(revenue-fuel cost)/hour / new engine cost*
Conway Mk 508	Boeing 707	1960	0·25%
Avon Mk 533	Caravelle 6	1960	0·27%
Spey Mk 512	BAC 1–11	1965	0·39%
RB 211–22	Lockheed Tristar	1972	0·43% (estimated data)

(This assumes revenue at $2\frac{1}{2}$p/seat mile, 0·8 Mach Number cruise at 36070 ft and 90% of Maximum Continuous Net Thrust)

In comparison with this, the maximum unreliability cost during the initial in-service phase expressed as delay cost per hour per engine/new engine cost is of the order 0·010% to 0·013% for all the above engines. This assumes a very pessimistic 5 delays per 1000 engine departures and £2000 to £5000 per delay, depending on aircraft size.

This does not cover overhaul and maintenance costs and so does not represent the whole engine earning story. However, the figures show convincingly enough why an engine manufacturer and his customers, the airlines, are prepared to knowingly enter a period of lower reliability than is being achieved by current in-service aeroplanes.

Similarly the design and manufacturing team have to be prepared to change from a proven design or a well understood material in order to save weight and/or volume and produce a more efficient product in a competitive marketing situation.

The capital costs now involved in large airliners and their propulsion systems are such that it makes economic sense to improve

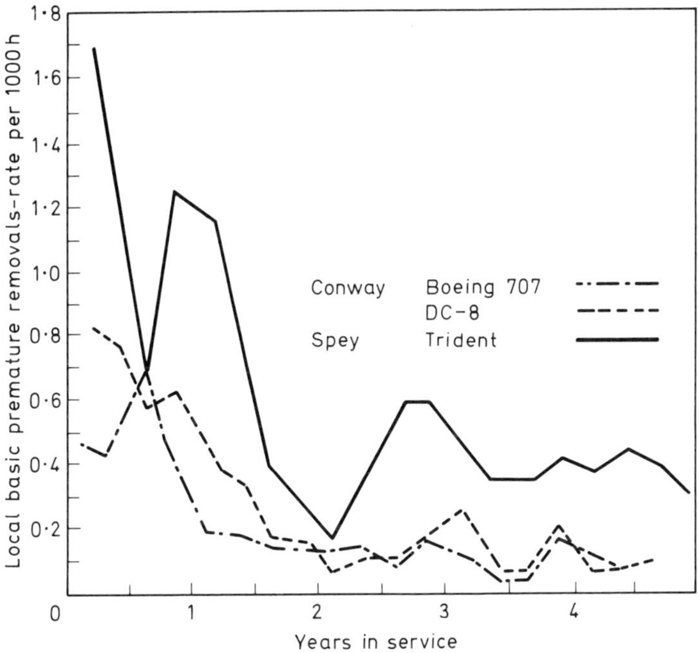

Figure 3.3 The improvement in the Reliability of a new product with the incorporation of modifications in the early in service period

the early reliability of the aircraft in the shortest possible time at some additional engineering expense. To give the reader some idea of the increasing cost to the customer arising out of any minor problem which produces a delay, Figure 3.4, compares the delay cost of an early turbo-prop with that of a large airbus.

This is just one example taken from an industry of a highly technical nature. The financial incentive to strive for higher reliability can be equally great in less technical industries. In a marketing situation where the customer is concerned only with first cost, the urge to aim for high reliability can be no more than a gentle nudge and a paragraph of lip service. The customer must be interested in *whole life costing* or else be concerned about the inconvenience or hazard of low reliability. Whole life costing, or *cost of ownership*, takes into account not only the initial price of the piece of equipment, but also the cost of operating it and keeping it in good repair by curing problems, replacing failed components, routine and non-routine maintenance, necessity or otherwise to have standby equipment, repair facilities and repair teams, all in addition to direct

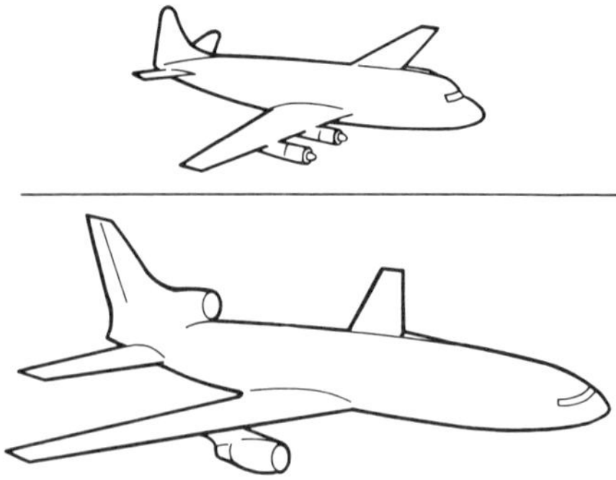

Revenue earning capacity per hour (approx)	Cost of 30 minute delay
£400	£210
£5000	£1250

Approximate prices in 1968

Figure 3.4 The increasing cost of unreliability with the increasing cost of capital equipment

operating costs and depreciation. If cash flows are taken into account this may reduce the case for accepting a higher initial cost in the interest of greater reliability, but on the other hand may increase the interest in securing fixed repair cost agreements. The latter can provide considerable incentives to a manufacturer to improve reliability. The pressure to change from a known reliable design or system is often due to the attraction of reduced operating costs or increased convenience. A solid fuel central heating system is a very reliable system indeed with attractive and competitive direct operating costs. However, the increased convenience attraction of automatic gas or oil fired systems does not seem to have been reduced by the attendant increase in unreliability. This is a more homely example where the attractions of increased convenience to the owner have overcome the potential annoyance of repair costs and reduced reliability. The advertisements for a number of these systems offer fixed cost maintenance schemes and show comparative figures on the total cost of ownership over a period of years. If the householder can understand and accept the concept of *total cost of ownership*, it seems a little strange that sometimes these ideas seem novel to parts of industry or departments of government.

Summing up this section, the excuses and reasons for changing from a known reliable design or manufacturing method must be based on sound financial arguments and not that some technical man is keen to try it a different way to please his technical ego.

3.3 MAKING INFORMATION AVAILABLE TO THE DESIGNER

Taking up the theme of 'what can we do that we have not done already?' let us first look at the design stage of a project. The designer has the job of making the best possible compromise to achieve a particular end. Whether he likes it or not he will be given plenty of conflicting advice from specialist engineers with regard to the requirements of performance, weight, material, manufacturing, inspection, and maintainability, but it is unlikely that he will have quantitative information on the reliability of possible alternative design configurations. Some of the additional effort should be in providing just this information because otherwise one is back to the old system of relying on memory and few changes in design personnel. In practice, memories appear to be fallible and personnel changes quite frequent.

One way to try to overcome changes of personnel and the unreliability of human memories is to catalogue design experience in some way. The list of modifications to a product or process can act as an index to the major changes in design, production methods, or perhaps inspection methods, but this is not sufficient in itself. The engineer on the project needs some brief handy reference to show visually the cures and modifications that have been tried in the past. When the redesign is needed or the process needs to be changed the pressure will be to provide a solution quickly, and typically time will not be available for extensive research and file searching to find a few relevant facts about the history of the part in question. It is not unknown for the same 'cure' to be produced ten years after its original use. A change is not necessarily for the better, and a history of the changes to date can improve the chances of not repeating an ineffective cure—and perhaps more important—can improve the chances of adopting a successful cure on a second or third occasion.

One large firm that has tackled this problem is Pratt and Whitney[13] who claim to have established a Design Co-ordination Manual which contains reports summarising the service history of specific problems, the particular designs involved and the effects of any corrective action taken. The purpose of the manual is to alert the Design Engineer to past problems and their solutions. The efficiency of the cure is included so that a series of cures may be detailed as well as the one that finally solved the problem.

Such a manual makes available to any designer the accumulated experience of many of his contemporaries and predecessors, and so can play a valuable part in achieving high product reliability. It does of course imply some investment before pay-off will occur, and a consistent policy will be required over a period of time if a design co-ordination manual is to become sufficiently comprehensive and cost effective. To start to assemble one and then to lose heart half way through is a complete waste of the initial investment.

3.4 SWALLOW YOUR PRIDE AND START WALKING

The next dodge to try when the design or redesign is still on the drawing board is the swallow your pride and start walking approach. This is a busy man's approach to failure analysis and to design review meetings (of which more is in the next chapter). This is not meant to be a criticism of the concept of design review, but it is a time consuming business and often takes a large elapsed time to complete because of the long period of notice required to collect together all the interested parties. In the swallow your pride and start walking approach, the engineer carrying out the design must forget that he is the most brilliant man in the business (which he may be) and set off with a print of his design schemes to tour the interested parties. The additional motivation that follows an engineer being approached for a quick opinion when the design is still flexible is something worth striving for.

Once a design scheme has been issued then a change becomes a battle for supremacy between the designer who is fighting to have his design accepted and the critic who can see some flaw in the proposed design.

If the swallow your pride and start walking approach has been adopted by the designer then he is almost fireproof against criticism. The critic is admitting his own shortcomings if, after having been offered a chance to alter the design, he comes along wanting change. Most of the improvements in reliability are gained from straightforward engineering commonsense and by offering a number of experienced engineers the opportunity to feed their experience into a design.

One of the rules of the game is to kickoff with the person whom you believe is least likely to like the design solution being put forward. It is often he, or she, who has the biggest contribution to make before the design is frozen. Or she? Yes, or she! The arrogant male can learn much from the female who may be less interested in the politics of her career and therefore less hesitant in pointing out that it is rather silly to do it that way, when all that is required is a simple alternative.

The Service Engineer is a must during the tour since it may be that the designer's previous efforts are reasonably reliable in service in spite of his efforts and not because of them. The customer may have found some maintenance dodge which keeps him out of trouble, but which would not be known to the next customer for the product. For example, in a less developed country where the equipment will be used in a less skilful manner and operated exactly as the instruction book says, catastrophe could occur. The Service Engineer may not understand the finer details of the technical input that has contributed to the design, but his practical approach to the less technical considerations may prevent a howler being sent off to the customer.

The other very practical input to the design must come from the Production Engineer. A reliable concept can be made invalid if the method of producing the hardware cannot produce a consistent job. The fact that one sample of the product is very reliable does not impress the customer who has a second sample which is unreliable. He tends to remember only the latter when he next orders new equipment. It is very tempting for the designer to grasp at each new production technique which gives him greater freedom to design new shapes or fastenings and produce a more elegant design. The Reliability Engineer should be suspicious of all this new nonsense until a suitable amount of test data is available, and the production technique has been well proved. There is of course a gleeful tendency for the Production Engineer to grasp at any opportunity to try out his new toy, and interrogation by the Designer can help to spot this before committing his design to a particular method of production.

3.5 RELIABILITY AND MAINTAINABILITY CHECKLISTS

In talking to the Designer in a firm which issued Designer's Checklists I asked a senior designer if I could see his checklist. He informed me that he had filed it in the waste bin and all the checklists he needed were in his head. A little arrogant perhaps, but that will be the fate of any checklist which is either too detailed, and therefore too long, or else too simple and therefore too short. A checklist with a specific purpose in mind can be used successfully as long as it is credible.

The most credible checklist of all is one that lists all the failure modes that have occurred on previous occasions, and if possible the rate at which the failures occurred in service. Such a list by definition cannot be ignored since it exposes all the major shortcomings of previous designs, and to consign such a list to the bin means that hard won previous experience is being consciously thrown away.

One very experienced designer of my acquaintance favoured sanctions and disciplinary action against the perpetrator of any design deficiency which could reasonably have been eliminated by careful attention to the checklist provided.

An extension to a checklist based on service failures would be the inclusion of those failure modes found on previous occasions and cured before the product reached the customer. This information is more difficult to correlate because often the standard of the equipment being tested on development is changing continuously and comparisons between tests may not be valid. Only where the test is not being aimed at producing failures is it fair to consider the failure modes as being genuine candidates for inclusion on a reliability checklist.

The psychology of this form of checklist is very apt, for there is a common belief among the designing fraternity that their designs will not fail. A permanent and authentic record of the failures of their predecessors will help combat this common belief. Equally well, if the checklist of previous failure modes is complete, then failure modes that do not appear on the list should not appear on the next design carried out to meet similar design criteria.

A credible maintainability checklist is also best compiled from reported difficulties that have occurred in service in the past. A car on the market about ten years ago had a grease nipple on the gear linkage which the manual stated should be greased at every major service. This was sound advice as far as the gear linkage was concerned but the problem was that the engine and gearbox had to be removed to get at the nipple. This particular problem was overcome by deleting the greasing operation from the maintenance schedule. Fortunately for the manufacturers this particular operation was a luxury and was not absolutely necessary for trouble free reliable operation of the vehicle. If the checklist had merely said: 'are all grease nipples accessible to a grease gun?' then the designer would tend to mentally say 'what a stupid thing to ask a fully grown designer'. If on the other hand the checklist is framed in the following fashion, it is likely to be given more than a cursory glance.

maintainability fault	*occurence*	*number of occurences in last 10 years*
Inability to reach grease nipple with grease gun	Nipple on gear linkage of the 195X model XX only accessible after removing gearbox and engine	1

So let us have no more holier-than-thou checklists which end up as garbage, but rather, credible ones which command attention because they are based on indisputable shortcomings that have actually occurred on previous occasions.

3.6 'WE'VE HAD A FAILURE, SO WE HAD BETTER DESIGN A NEW ONE'

Absolute nonsense! There should be quantitative evidence that a new design is required and has a chance of being better:

(1) to be worth the cost of producing the redesign
(2) to attract the customers into buying a modification

The minimum evidence for a change to a product already in service is the answer to the questions: 'What proportion of the customer's problems, and operating costs will be solved if the redesign is a complete cure?' Table 5.1 on page 71 shows that a comparatively small number of problems (say 9) produce half of the total disruptions to the customer's operation, and these are the problems that must be solved first. The story on operating costs is usually very similar, and extending the lives and improving the reliability of a modest number of parts can reduce the operating costs by a large amount. It sounds terribly obvious, but suppose that the trouble is distributed as follows:

problem No.	*% of total trouble to operator*
1	15%
2	12%
3	9%
4	9%
5	4%
6	3%
7	2%
8—76	46%
	100%

and the available engineering manpower is a team of 10, then a sensible distribution of effort is:

problem No.	*No. of engineers on problem*
1	3
2	2
3	2
4	2
5, 6, 7, 8—76	1

The fact that problems numbers 9, 12, 16, 24, 36, 45, 54 are technically fascinating is of no interest to the customer. The only justification for leaving 10% of the effort on problems 5 onwards is to act as lookout man to make sure that an apparently insignificant problem is not about to develop rapidly into a major one. The main reasons for arguing thus are:

(1) financial (see reasoning above)
(2) statistical (read on)

Problem No. 5 is contributing 4% of the total number of failures. The significance of this depends upon the size of the operation, so let us assume:

200 units in operation
10 hours use per day
total failure rate = 0·2/1000 hours
time between overhaul = 2000 hours
The number of failures/year due to problem No. 5, on the average, is: $200 \times 10 \times 365 \times \frac{0{\cdot}2}{1000} \times \frac{4}{100} = 6$

This compares with the total number of scheduled interruptions for overhaul of: $\dfrac{200 \times 10 \times 365}{2000} = 365$

Assuming no safety problem, the interruptions due to problem No. 5 look pretty insignificant compared with the routine interruptions. The situation would only change if the time between overhauls (T.B.O.) were increased. If the T.B.O. increased to 20 000 hours, so that the scheduled interruptions became 36 per year, then the 6 unscheduled interruptions become very significant.

Consider now problem No. 20, which may be contributing 0·7% of the total trouble, or for the operation of 200 units being considered, about 1 failure per year. It would often be wrong to try to improve this situation because of the difficulty of proving that a new design would be any better, or even as good as the old. If only 1 failure per year is expected from the old design, it will be 2 years before there is much evidence that the new standard is any better. Will anyone want to buy it then anyway? If you cannot put a value to it it probably is not worth doing, and that is why it is worth analysing service data, of which more in Chapter Eight.

4

Failure analysis and mathematical models

4.1 THE CASE FOR FAILURE ANALYSIS

There are various types of analysis which may be carried out at the design stage, or when establishing the machining programme, which are used to establish where improvements can be made. Some of their titles are *Failure Mode and Effects Analyses, Reliability Analyses, Safety Analyses, Design Analyses* or *System Analyses.* The relevant literature contains a number of detailed instructions and considerable benefits are claimed.

The aim of all these types of analysis is to adopt a documented and disciplined approach to the particular design of a system, machine, component or process and to carefully (and unfortunately laboriously) work through it considering in turn the affects and consequences of each failure or malfunction which could occur. Typical form headings for carrying out this type of analysis are shown in Figure 4.1.

The main differences between a Reliability Analysis and a Failure Mode and Effects Analysis is that the former endeavours to make quantitative assessments of the in-service reliability associated with each failure mode. The latter may or may not include quantitative assessments as appropriate to the particular equipment, system or process being analysed. Any analysis should ensure that detailed attention is given to each component within an assembly, but the more critical components should be considered in detail. Similarly, in analysing a system, additional attention should be given to those parts of the system where failure produces the most unpleasant effects. For maximum credibility the defect modes considered should be based primarily on modes that have occurred on previous occasions, but modes which have not been experienced should be included. Any prediction on failure rates due to modes which have

1 *Sub-Assembly or Part Description*	2 *Identification Number*	3 *Quantity*	4 *Possible Defect Mode(s) and Degree of Severity*	5 *Possible Secondary Defect Modes and Degree of Severity*	6 *Product Service Data and Predicted Reliability of New Design*					7 *Effect on Sub-Assembly Operation*	8 *Effect on Complete System Operation if assessable*	9 *Effect on Engine* Operation if assessable*	10 *Proposed Modifications to rectify defect*	11 *Recognisable Symptoms of defect*	12 *Defect Classification*	13 *Contributory Remarks (including Data Source, Design Improvements and effect of changes of duty)*
					Total Population	*Total Operating hours*	*Total Confirmed Defects*	*Rate per 10^6 hours*	*Predicted Rate of New Design per 10^6 hours*							

(a) RELIABILITY ANALYSIS

Item No.	*Item*	*Failure Mode*	*Failure Mechanism*	*Effect of Failure*		*Failure Category*	*Indications*	*Preventative Measures taken on Design*	*Recommended Action*
				Primary	*Secondary*				

(b) FAILURE MODE AFFECT ANALYSIS

*machine, plant etc. as appropriate

Figure 4.1

not yet occurred must be somewhat speculative. The classification of defect modes into categories dependent upon the consequences of such defect modes being experienced, gives the analyst an indication of those areas to which particular attention should be paid.

The inclusion of quantitative assessments and estimates can be worthwhile even if the absolute numbers assigned to a particular mode are open to debate. They usually are. In deciding which action to take it is the relative importance of failure modes in terms of result and frequency that should set the priorities for attention. The absolute level of reliability computed by the addition of each individual failure rate in an analysis is open to question. Reliability estimates of electronic equipment or systems usually turn out to be pessimistic perhaps by a factor of ten. This in no way invalidates the comparison of two alternative designs and enables a decision to be reached based of factual quantitative data.

As mentioned earlier, and to be mentioned again later, the more a decision to take a particular route can be based on quantitative data, the more likely it will be that the decision reached is the best way.

Another advantage of quoting quantitative estimates of component, process or system reliability is that the analysis can be used as a dynamic document for monitoring improvement throughout an entire programme. The improvement or otherwise gained or lost by changes introduced during the programmes can be controlled and appreciated from the point of view of the effect of changes on equipment or process reliability. In a similar fashion the analysis can be used to assess the effect of a change in environment or duty by adjusting the quantitative assessments of those components or processes which will be affected by the proposed change. In this way a rational decision can be made on whether the change proposed will result in an acceptable or unacceptable reliability being achieved.

A typical reaction of a supplier when asked to provide such an analysis for his process or equipment is:

> 'It is all very well your asking me to carry out a failure mode and effect analysis, or a reliability analysis, but in the analysis you ask me to state the reliability of components (processes) and I have no data available. How can I start to predict anything?'

This reaction provides the clue to the main benefit of carrying out an analysis. It compels the analyst to search for and examine such information as may be available, and the experience gained from examinations of service trouble reports, spares inventories and orders, customer's complaints will highlight where the problems will arise and are likely to be located. Without the requirement to carry out an analysis it is less certain that these investigations into the quantitative behaviour of existing equipment or processes will be made. In a firm where adequate management information systems are already established and used then there is less advantage in performing a quantitative analysis. Are you convinced that in your firm the left hand knows what the right hand is doing?

The original designer or other type of engineer will have already carried out mental analyses to assess the possible consequences of a malfunction, but only in rare cases will this analysis have been committed to paper. There is an unfortunate tendency for design and production engineers to expect the best and to assume that unlikely events will not happen. However, we have already seen earlier that most phenomena are distributed in nature and can occur at any time. Even though the total number of occurrences may be few, they could occur sooner rather than later for the same overall failure or malfunction rate. Although this is widely appreciated it tends to remain a fact that is appreciated rather than a fact that produces action. Engineers in general are optimists who assume that their designs or processes will work all the time. None of this is true in a situation where there is a question of hazard or considerations of safety apply, but it is true in the majority of non hazardous engineering situations.

The aim of a formal analysis is to spot anything which was missed in the original concept. This formal analysis can be carried out by the engineer who originated the design and it may be argued that he is the ideal person. A counter argument is that he is so concerned and absorbed in the detail that something obvious, or less obvious, may have been missed. By adopting a *broad brush* approach it may well be that shortcomings in the design or process can be spotted before paper is transformed into hardware. The results of formal analyses may be more acceptable to the designer if performed by one of his peers rather than by an alleged superior or subordinate. Whilst this might be true, there is also a general tendency for like to stay with like and the amount of constructive criticism is likely to be greater from a non peer if the normal rules of competitive motivation apply.

Having been involved in the execution or co-ordination of a number of formal failure mode and effects analyses and reliability analyses, it is not easy to quantify the benefits with any precision. What happened in practice was that the formal analysis set up a train of thought in the original designer's mind and formed one of the inputs which may have caused him to think again and assess his original ideas with a more critical eye.

The recommendations made in formal analyses were often not accepted initially, but checking up a few months later on the subsequent actions that had taken place revealed that in a number of cases some of the changes recommended, but not accepted, had been made. In some cases this may be because the original concept was defended by its conceiver against initial attack, but a period of sober reflection resulted in the merits of the change being accepted. Alternatively it sometimes occurs that an additional piece of evidence, for example, from a rig test or from a pilot process run, suggests that the doubts raised in the formal analysis were in fact valid. Sometimes this additional evidence would be sufficient in itself to merit a change being made, but for a number of cases the issue is less clear and several inputs are required before the best course of action is clear and a correct decision made.

Contrary to the above discussion on peers, superiors and subordinates, a considerable number of failure analyses have been performed in a competent fashion using trainees. At first sight this looks a bit strange, but it has been found that once their initial shyness and reserve have been conquered, an intelligent trainee will query and challenge concepts and ideas which have been long accepted as the normal way of doing things. This sort of interaction can be very healthy and sobering to an engineer who thinks he knows it all and accepts established methods without question. The success of using trainee labour is only due to the co-operation received from the original designer and therefore it is vital to persuade him, before embarking upon a formal analysis of his work, that the venture is worthwhile.

Once the engineer realises that the whole purpose of the exercise is to improve the quality of the output on which he is ultimately judged, then the exercise will normally progress unimpeded by politics.

There is no substitute for actually performing a failure mode and effect analysis in trying to assess whether this approach may be useful in your particular situation or industry. It is a job, and not usually a very popular job, for those who are highly motivated towards achieving excellence in their work. It is recommended that you put your hand into your inside pocket, pull out a pen or pencil and make a start at analysing the failure modes and effects of the following control system. Once you have done that, and if you believe that this approach may have something to contribute to your situation then try out a pilot exercise for a design or process in your area and see where you get to.

4.2 YOUR CHANCE TO TRY FAILURE ANALYSIS

Background information to the control system illustrated (Figure 4.2) is that an exothermic reaction is to be controlled by using a

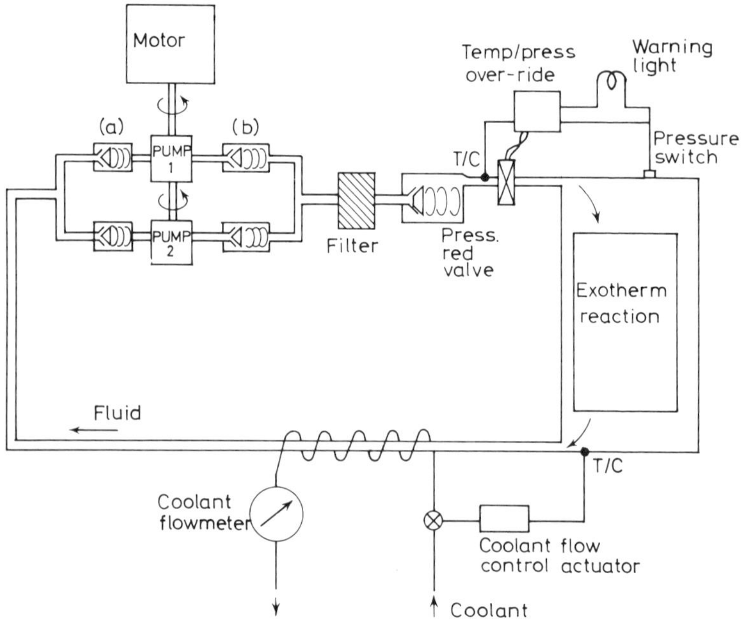

Figure 4.2 Reaction Control System

primary coolant fluid in a closed circuit and the primary coolant is cooled by a secondary coolant in an open circuit. Two pumps circulate the primary fluid at controlled temperatures and pressures to control the conditions in the reaction vessel. Should the temperature or pressure go beyond the design limits then the reaction may become unstable and the possibility of an explosion arises. The primary fluid is pumped through a further pressure reducing valve and through a temperature and pressure override shut-off valve before entering the reaction vessel. Primary fluid temperature is controlled by sensing reaction vessel temperature at coolant exit and controlling the flow of the secondary coolant through the heat exchanger to maintain the primary coolant exit temperature constant. Nonreturn valves are used to isolate the two pumps, the design concept being that dual pumps provide for redundancy and safe operation in the event of a single pump failure. The temperature and pressure override valve is actuated by a signal from either a thermocouple at reaction vessel entry or by a pressure gauge in the vessel itself. In the event of the override operating a warning light is lit.

Table 4.1 FAILURE MODE AND EFFECTS ANALYSIS

Item No.	*Item*	*Failure Mode*	*Failure Mechanism*	*Effect of Failure*		*Failure Category*	*Indications*	*Design Preventative Measures*	*Recommended Action*
				Primary	*Secondary*				
1	Motor	1.1 Fails to rotate	(a) Power failure (b) Circuit braker out (c) Windings burn out (d) Seizure	No fluid pumped	Reaction out of control, possible explosion.	H			Provide duplicate motors, separate power supplies
		1.2 Overspeeds	(e) Control gear failure	High pump delivery pressure	Pressure reducing valve moves to nearly closed, later pressure switch closes off.	N or I	Red light warns operator		Provide audible warning, e.g. hooter.
		1.3 Runs slow	(f) Control gear failure (g) Partial seizure (h) Partial burn out	Less fluid pumped	Reaction gets warm, coolant flow goes up	N or I	Flowmeter warns operator		Visible or audible warning on flowmeter. Trend analysis on flowmeter.
2	Pump 1 or Pump 2	2.1 Fails to pump	(a) Drive fails to pump 2	No fluid pumped from Pump 2	Non-return Valve closes other pump continues	I		Duplicate pumps provided	
			(b) Outlet pipe fractures	Some fluid lost		H	Fluid leaks	Conservative stresses	Pressure test when commissioning
			(c) Seizes or jams or drive to pump 1 fails	No fluid pumped from pump 1 or 2	Reaction out of control Explosion	H			Provide separate motors and drives
2	Pump 1 and Pump 2	2.2 Low delivery flows	(a) Partial seizure of drive on pump (b) Wear in pump	Less fluid pumped	As 1.3	N or I	As 1.3	Duplicate pumps at low loading avoid seizure and wear	As 1.3

Failure Categories: H Hazardous; I Inconvenience – unscheduled maintenance required; N Negligible effect.

The first sheet of the analysis table has been completed as shown in Table 4.1 but the remaining work awaits the sharpness of your mind and pencil. The analysis when performed should convince you that the control system of Figure 4.2 is the product of an inadequate designer. Your analysis should highlight the worst and least reliable features of the design in a convincing and carefully documented way. Other literature suggests even more painstaking analyses that can be used for greater reliability.

4.3 RELIABILITY LOGIC AND MATHEMATICAL MODELS

This is a long winded title for something that is rather simple. It is all based on:

(a) Reliability of components in series

This is the 'fairy lights on the Christmas tree' case (Figure 4.3).

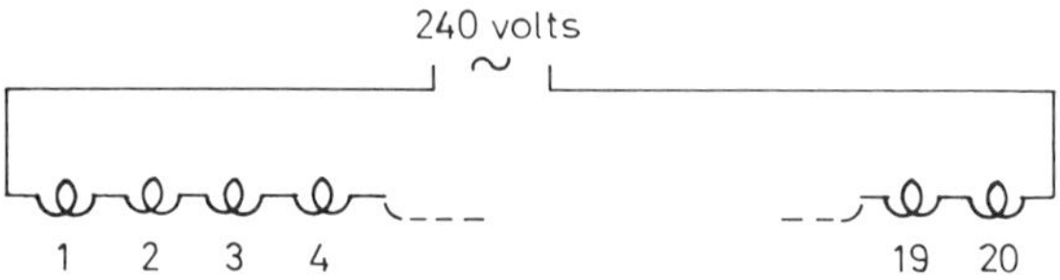

Figure 4.3 Reliability of Components in Series

Twenty 12V bulbs are connected in series and run off the mains. If one fails, all the lights go out, and there are twenty items that can fail.

System Reliability R_s = Reliability of lamp 1 × Reliability of lamp 2 × ....Reliability of lamp 20

or

$$R_s = R_1 R_2 R_3 R_4 R_5 R_6 R_7 R_8 R_9 R_{10} R_{11} \dots R_{20}$$

For n identical components in series:

$$R_s = (R_c)^n$$

(b) Reliability of components in parallel

This is the 'wall lights' case (Figure 4.4).

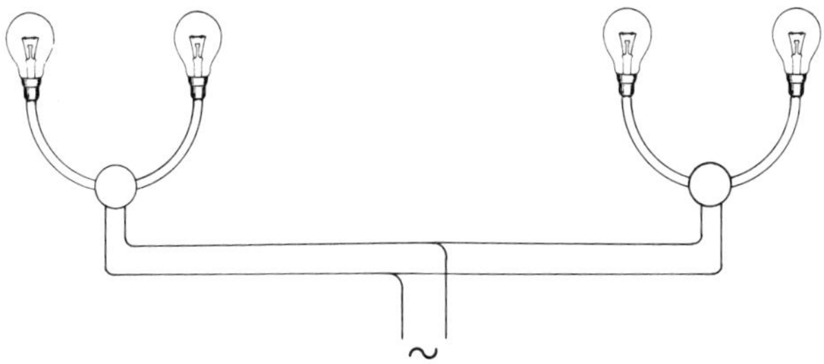

Figure 4.4 Reliability of Components in Parallel

Four lights are all in parallel and if one bulb fails the other three continue to work.

$$\text{System Reliability } R_s = 1 - \text{Unreliability of total}$$
$$1 - Q_t$$

$$Q_t = \text{Unreliability of lamp } 1 \times \text{Unreliability of lamp } 2 \times \ \dots \ \text{lamp } 4$$
$$= Q_1Q_2Q_3Q_4$$

For n identical components in parallel, where all must fail for system failure:

$$R_s = 1 - Q^n$$

(c) Binomial Law of Reliability

This law states that when n equal components are in parallel, and R and Q are the reliability and the unreliability of each component then:

$$(R+Q)^n = 1$$

which may be expanded:

$$R^n + nR^{n-1}Q + \frac{n(n-1)}{(2)(1)}R^{n-2}Q^2 + \dots {}^nCrR^{n-r}Q^r + \dots Q^n = 1$$

If *all n* components must fail to produce failure (all wall lights go out to produce darkness).

$$R_t = 1 - Q^n$$

If only r components must fail to produce total failure (a three engined aeroplane cannot take off fully laden with two engines out),

$$R_t = R^n + nR^{n-1}Q + \ldots + {}^nCr - 1R^{n-(r-1)}Q^{r-1}$$

and since $R_t = 1 - Q_t$

$$Q_t = {}^nCrR^{n-r}Q^r + \ldots + Q^n$$

This is probably better understood with a simple example. Consider three generators in service, at least one of which must operate to provide the necessary power (Figure 4.5).

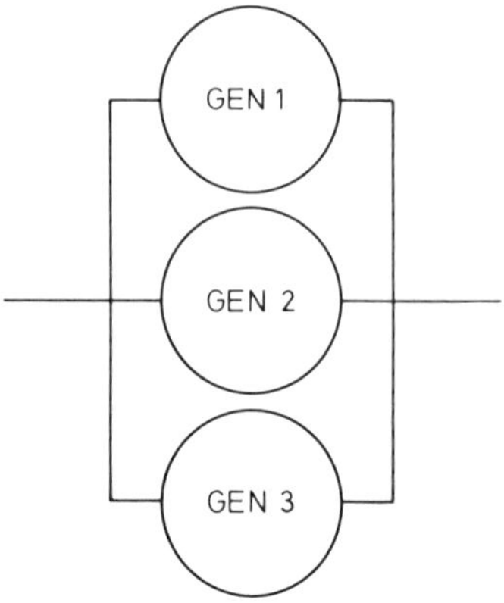

Figure 4.5 Three generators operating in parallel

Let Reliability of any Generator = 0·9
∴ Unreliability of any Generator = 0·1

$$R_t = R^3 + 3R^2Q + 3RQ^2$$
$$= 0{\cdot}729 + 3(0{\cdot}81)(0{\cdot}1) + 3(0{\cdot}9)(0{\cdot}01)$$
$$= 0{\cdot}999$$
$$Q_t = Q^3 = 0{\cdot}1^3 = 0{\cdot}001$$

and $R_t + Q_t = 1$

Where the individual reliabilities of components in a system are different, and possibly there are several failure modes, then there is no short cut by using the binomial expansion and each failure possible should be written down.

(d) Exponential Distribution

This represents the constant failure rate case where the equipment failure rate is independent of age.

Let $F(t)$ = total number of failures at time t, then:

$$\frac{dF(t)}{dt} = n\lambda$$

where n = number exposed at time t
and λ = constant failure rate of each unit in the sample n

If N = number originally exposed, i.e. $N - F(t) = n$

$$\frac{dF(t)}{n} = \frac{dF(t)}{N - F(t)} = \lambda dt$$

$$\therefore -\log [N - F(t)] = \lambda t + \text{const } (C)$$

$F(t) = 0$ at $t = 0$,

$$\therefore -\log N = \text{const } (C)$$

$$\therefore \log \frac{N}{N - F(t)} = \lambda t$$

$$\frac{N}{N - F(t)} = \exp (\lambda t)$$

$$1 - \frac{F(t)}{N} = \exp (-\lambda t)$$

$$F(t) = N[1 - \exp (-\lambda t)]$$

This is plotted in Figure 4.6.

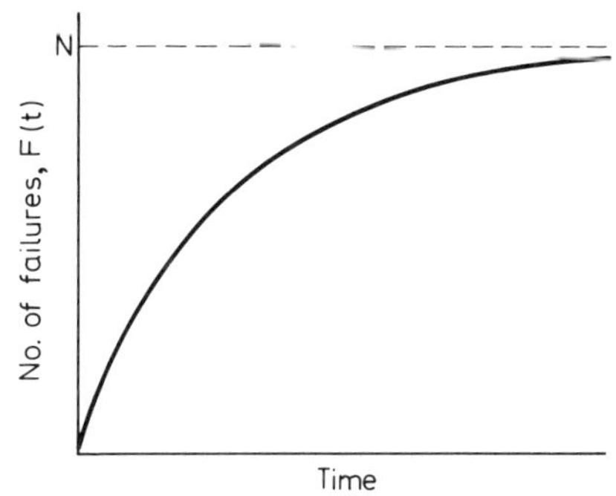

Figure 4.6 Exponential increase in the number of failures

$$\text{Unreliability } Q = \frac{F(t)}{N} = \frac{N(1 - \exp (-\lambda t))}{N} = 1 - \exp (-\lambda t)$$

$$R = 1 - Q = \exp (-\lambda t)$$

Hence, for the example in Figure 4.7:

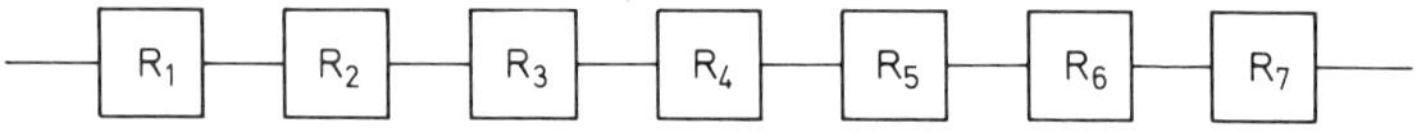

Figure 4.7 Simple series system

$$R = R_1R_2R_3R_4R_5R_6R_7$$

$$R = \exp-(\lambda_1+\lambda_2+\lambda_3+\lambda_4+\lambda_5+\lambda_6+\lambda_7)t$$

consider

$$\lambda_1 = \lambda_2 = \lambda_3 = \lambda_4 = \lambda_5 = \lambda_6 = \lambda_7 = \frac{1}{1000}\left(\frac{\text{failure}}{\text{hours}}\right)$$

Case 1 Exposure time t = 100 hours

$$R = \exp\frac{-7}{1000}\times 100 = \exp-0{\cdot}7 = 0{\cdot}496$$

or

$$R = R_1R_2R_3R_4R_5R_6R_7$$

$$= (1-\frac{1}{1000}\times 100)^7 = 0{\cdot}9^7 = 0{\cdot}48$$

Case 2 Exposure time t = 10 hours

$$R = \exp\frac{-7}{1000}\times 10 = \exp-(0{\cdot}07) = 0{\cdot}9325$$

or

$$R = (1-\frac{1}{1000}\times 10)^7 = 0{\cdot}99^7 = 0{\cdot}932$$

The two methods only give answers very close to each other if the individual reliabilities are high, because the assumption is made that this is the failure rate assuming everything else works. λ is really therefore a little smaller than the value used, so the $R = \exp(-\lambda t)$ sum is slightly optimistic.

It is important to understand and to be able to complete the elementary block diagrams so that reliability of alternative concepts can be settled quickly by quantitative assessments.

There are some simple examples to try in Figures 4.8 to 4.11. The answers are at the back of the book.

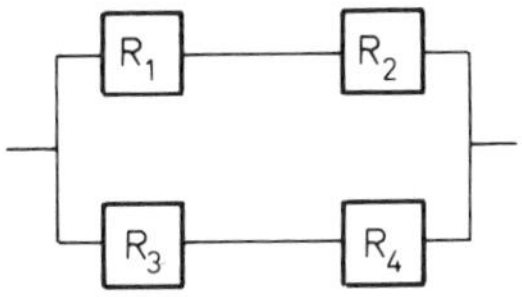

Figure 4.8 Series-Parallel system

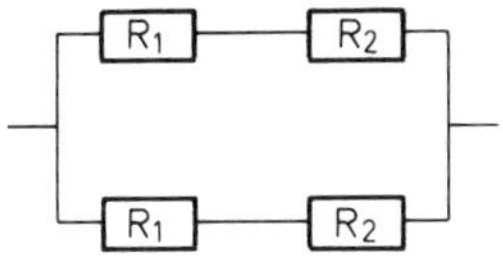

Figure 4.9 Series-Parallel system with identical paths

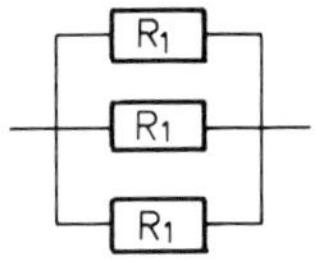

Figure 4.10 Triplicated system

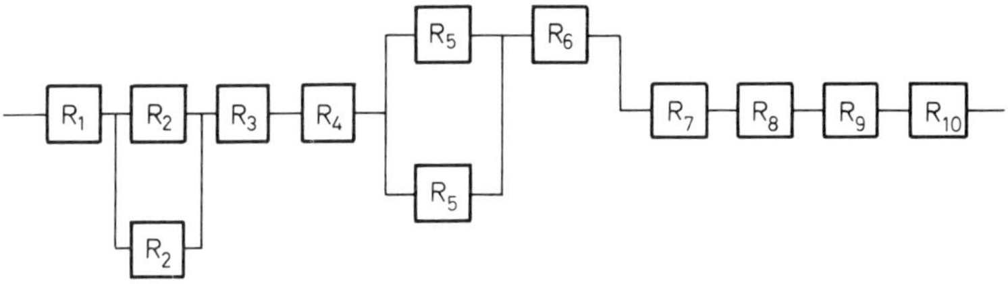

Figure 4.11 Series system with some Stand-by parallel components

4.4 RELIABILITY LOGIC WHEN YOU HAVE NO DATA

It is possible on some occasions to decide quantitatively between alternative design ideas when little is known about the reliability of individual components. This sounds a somewhat dubious claim, but a study of the following example may give some credence to the idea of doing sums without data.

Consider two signal generators. The designer is trying to decide between (a) 3—20 AWG wires or (b) 4—22 AWG wires. Ideally he would like to use 4—18 AWG or 4—20 AWG wires for simplicity and reliability, but there is only room for (a) or (b). To make matters worse, he does not know much about the reliability of the different wire thicknesses. These two alternative approaches are shown in Figure 4.12.

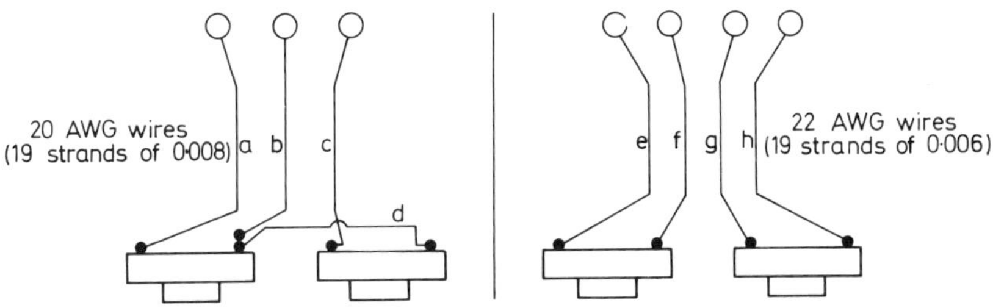

Figure 4.12 Alternative wiring configurations using three thicker wires or four thinner wires

With system failure defined as in Figure 4.12:

THREE WIRE SYSTEM		FOUR WIRE SYSTEM	
Failure Combination	Probability of Occurrence	Failure Combination	Probability of Occurence
Open circuit failures which will cause system failures are:			
a and c	P_{O1}^2	e and g	P_{O2}^2
a and d	P_{O1}^2	e and h	P_{O2}^2
b	P_{O1}	f and g	P_{O2}^2
		f and h	P_{O2}^2
Short circuit failures between wires causing system failure are:			
a to b and b to c	P_{S1}^2	e to f and g to h	P_{S2}^2
a to b and c to d	P_{S1}^2	e to h and f to g	P_{S2}^2
a to b and a to c	P_{S1}^2	e to g and f to h	P_{S2}^2
a to c and b to c	P_{S1}^2	e to g to h to f	P_{S2}^3
a to c and c to d	P_{S1}^2	g to e to f to h	P_{S2}^3
		e to h to g to f	P_{S2}^3
		h to e to f to g	P_{S2}^3
Short circuit failures to earth causing system failure are:			
a, b and c to earth	P_{E1}^3	e, f, g, h, to earth	P_{E2}^4
Total: $P_{O1}+2P_{O1}^2+5P_{S1}^2+P_{E1}^3$		$4P_{O2}^2+3P_{S2}^2+4P_{S2}^3+P_{E2}^4$	

Values for each of the four probabilities are unknown. However, if we assume that the probability of open circuit failure is equal to the probability of short circuit failure to earth, we have:

$$P_{O1} = P_{E1} = P_1 \text{ (say) and } P_{O2} = P_{E2} = P_2 \text{ (say)}$$

Since short circuit failures between wires demand the failure of two separate thicknesses of insulation, this failure is liable to take twice as long as a short circuit to earth and its probability of occurrence in a given period will be approximately half the probability of a short to earth. This means:

$$P_{S1} = \tfrac{1}{2}P_{E1} = \tfrac{1}{2}P_1 \text{ and } P_{S2} = \tfrac{1}{2}P_{E2} = \tfrac{1}{2}P_2$$

Therefore the probability of system (1) failing becomes

$$P_1+2P_1^2+5(\tfrac{1}{2}P_1)^2+P_1^3 = P_1+\tfrac{13}{4}P_1^2+P_1^3$$

and the probability of system (2) failing becomes

$$4P_2^2+3(\tfrac{1}{2}P_2)^2+4(\tfrac{1}{2}P_2)^3+P_2^4 = \tfrac{19}{4}P_2^2+\tfrac{1}{2}P_2^3+P_2^4$$

If we assume $P_1 = 10^{-5}$ (ten defects per million hours), the probability of failure of (1) is

$$\frac{1}{10^5}+\frac{13}{4\times 10^{10}} \approx \frac{1}{10^5}$$

Therefore if the alternatives are to have equal reliability:

$$\tfrac{19}{4}P_2^2+\tfrac{1}{2}P_2^3+P_2^4 = \frac{1}{10^5}$$

P_2 being small ($\approx 10^5$), P_2^3 and P_2^4 are negligible, therefore:

$$\tfrac{19}{4}P_2^2 \approx \frac{1}{10^5}$$

$$P_2 = \left(\frac{4}{19\times 10^5}\right)^{1/2} = \frac{2}{10^2}\left(\frac{1}{1{\cdot}9}\right)^{1/2} = \frac{145}{10^5}$$

that is the probability of failure of the thinner wires in proposal (2) must be approximately 145 times more likely than the probability of failure of those in proposal (1) for equal probability of system failure. This is very hard to believe, and so proposal (2) is the correct one to select. I know of designers arguing for hours on similar topics, when just a quarter of an hour's sums would resolve the choice.

As has been said before, and will be said again, if quantitative data can be used to assess alternatives, the decision is easier to make. It takes away the emotion and prejudice that frequently comes out when a choice has to be made. This is only true before the decision logic is laid down. For instance, system 2 is the more attractive solution from the point of view of manufacture, but if system 1 looks more than twice as reliable it will be chosen.

It is a waste of time to examine the Reliability Logic if the Decision Logic is not laid down. For instance, 'I cannot decide which system to choose, so I will examine their comparative reliabilities, and think about it'. This is another way of saying that nobody wants to make a decision. Properly used, Reliability Logic can make the choice between alternatives much easier.

5

The development stage and designing test programmes

5.1 WHAT IS DEVELOPMENT ALL ABOUT?

The designer tries his best to produce a design that will meet every requirement in the specification at a particular price, and often in a timescale shorter than he believes reasonable. There are still instances in technical industries where equipment is designed, manufactured, and put into operation with only a short pass-off test. Two spectacular examples of this according to newspaper reports were the Q.E.2 and the Ronan Point Flats. At the other end of the scale according to newspaper reports, there is the Concorde Supersonic Transport Aircraft and Apollo space vehicles, where the considerable expensive amount of development testing carried out is obvious to the layman.

In the latter two cases, and in many other cases, testing is carried out after initial design and before delivery to the customer. This is the development test phase, when the aim is to convert a new, *as designed* product into a reliable piece of equipment which meets the specification and can be sold to the customer as a going proposition.

Before we all start feeling too self-satisfied that our own particular Company develops its products before selling them, there is still a tendency to design test rigs and expect *them* to work first time without any development testing at all. This is being just as naïve as expecting a product to work reliably the first time it is made. An undeveloped testing rig can ruin a complete development programme and cause long and expensive delays. Nearly all designs require some development. Development testing costs money and takes time. If the cost is too great, the profitability of the project may be jeopardised. If the timescale is too great, competitors may scoop the market, and again the profitability of the project is at risk. In an era of inflation, extended timescales inevitably mean increased total development costs.

Development is the art/science of finding the weaknesses and shortcomings of a design at a cost and in a timescale appropriate to the market situation. It could indeed be claimed that if all development were a science the product would never be launched in an acceptable timescale. There is never, repeat never enough time or money to run samples of equipment at the normal operating conditions until all the weaknesses have been detected. No decision to change a design to improve its reliability should be made until a weakness is highlighted, and this point is nicely made by Nixon in his *Fallacy of the type test* diagram (Figure 5.1).

Both design A and B will pass the type test, so no decision can be made at that point in time to say which design is preferable. If the

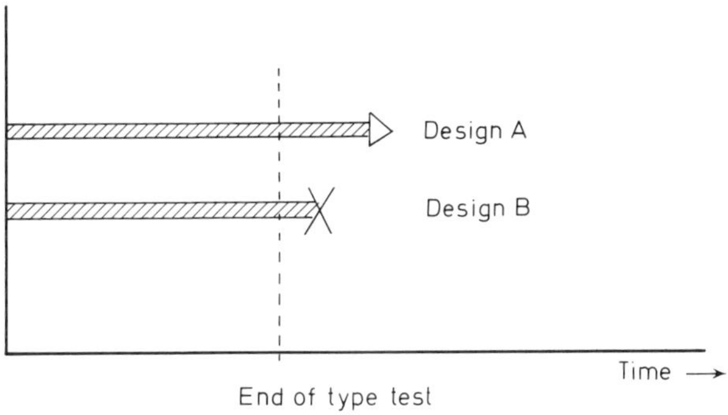

Figure 5.1 Fallacy of the Type Test

test is continued to failure and design B fails first then design A is the correct choice. Although it is desirable that development tests show that the service life will be satisfactory, the more important purpose is to highlight the inherent weaknesses in the product. To do this there is a need to test not only at the normal working stresses, but at *overstress* or *killing* test levels.

In overstress tests, the equipment is operated outside the design specification by increasing the severity of *all* the relevant degrading parameters and finding out which component breaks first. Although scientific training tells us to hold constant all but one variable, Shainin[14] explains how this is at best inefficient, and at worst misleading when applied to the development process.

The benefits of overstress testing are most easily appreciated by considering the cyclic fatigue failure case of Figure 5.2. If the test is carried out at the normal working stress, 10^5 cycles are needed to show that design A is inferior to design B. If the test is carried out at the overstress shown, then only 10^4 cycles ($\frac{1}{10}$ of the previous time and cost) are needed to show the same thing. The overstress must be to a reasonable level only, otherwise the failure mode may be totally unrealistic for the service situation. If the environmental temperature is also raised the wrong failure mode may be produced, such as creep failure in a metallic component instead of fatigue failure, or cavitation instead of corrosion in a hydraulic system. The failure mode of an egg normally subjected to shock loading at 15°C compared with the failure mode when tested at an overstress temperature of 100°C are particularly interesting and make this point rather nicely.

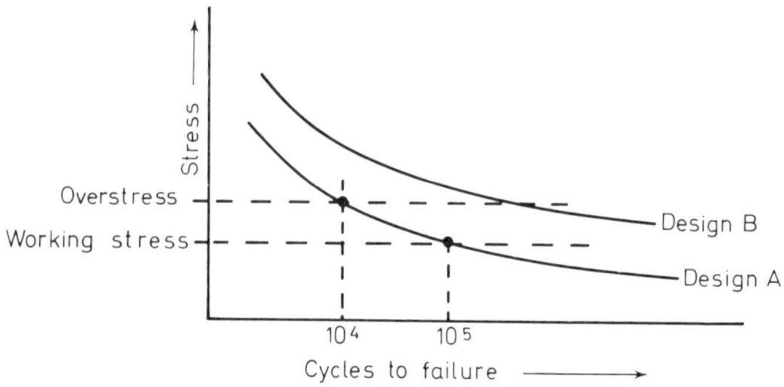

Figure 5.2 Overstress testing to reduce testing time

Stress should be thought of not only as applied load, but also as any relevant environment such as temperature, pressure, vibration, corrosive atmosphere, adverse tolerances, inferior surface finish, voltage fluctuation, dirty hydraulic fluids, mishandling or cyclic operation. Not all these will apply in every case, but in writing a development test schedule, engineering judgement must be used to select the most appropriate overstress factors in order to produce the failure modes which might occur in service.

A modern example of the failure of the scientific approach with an electronic transducer is shown in Figure 5.3. The manufacturer had cycled the transducer in the laboratory for an advisably large number of cycles, and overstressed the range of signals that were fed to the transformer. The transducer passed these extended overstress reliability tests with flying colours.

When installed on the prime mover that was to be monitored by the transducer, the inflexible special flex attached to the transducer failed repeatedly and rapidly. A few pulls on the wire during the lengthy laboratory testing would have shown up this fault before the customer received the transducer. The Development Engineer

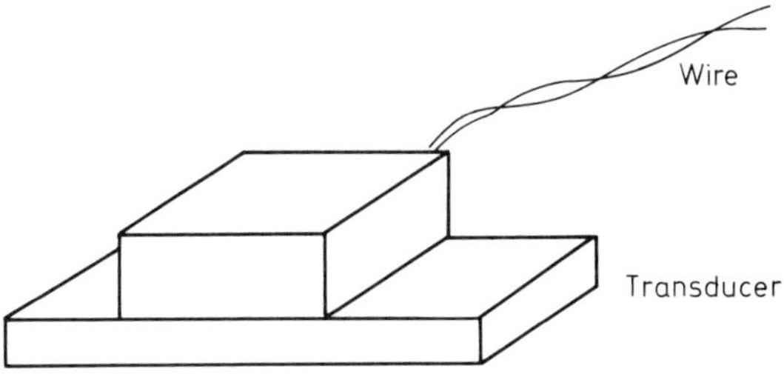

Figure 5.3 Transducer to be subjected to reliability test

had not included sufficient overstress factors for the overstress test to be any use at all in showing what parts of the product were going to cause unreliable operation in service. The science was good but the test was poor.

5.2 DESIGNING TEST PROGRAMMES

The classical way to design a test programme is to hold constant all the variables except for one. This is an excellent way to establish the influence of that one variable on the performance of the equipment, but it is also an inefficient method. For example the performance of a piece of equipment may depend upon three major variables and the traditional way to assess the influence of each is to hold constant two of them and vary the third. Most cautious engineers would also want to repeat the test more than once to ensure that their results were repeatable. They know about coin tossing and the laws of chance and therefore they want to know that their first test is not the 1 in 20 result that gives a misleading answer. This means that to assess the effect of two values of the variable, four tests are required.

More efficient systems of testing have been publicised by Shainin[14] who points out the tremendous improvement in efficiency in terms of information gained from a series of tests if the factorial method of testing is adopted. While the reader is recommended to read Shainin's explanation in full before embarking on the design of a major development programme, some of his ideas are included in the following sections.

Pre-planned tests and factorial testing

The usual objection to changing more than one variable is that it is not possible to say which variable is the important one in producing the desired improvement. It may in fact be a combination of variables producing the desired improvement and changing only one will not allow this state of affairs to be investigated. It is not normally possible to sort out the variable after the test and pre-test planning is necessary. The simplest case is for two variables, A and B. In a complicated structure it is reasonable to consider:

A is the material; $A+$ being a high strength expensive material
$A-$ is the standard material
B is the finishing process: $B+$ is a dressed weld
$B-$ is an undressed weld

First cost consideration makes $A+B+$ an undesirable combination, but to see if the life is superior, a cyclic test of the following combinations could be carried out at an overstress to give failure in 1/10th of the desired service life:

	$A+$	$A-$
$B+$	L_1	L_2
$B-$	L_3	L_4

L_1, L_2, L_3, L_4 represent the outcome of the test and are the achieved lives at the overstress test condition. The effect of the material change is therefore the difference between:

$$\tfrac{1}{2}(L_1+L_3) \text{ and } \tfrac{1}{2}(L_2+L_4)$$

The effect of the dressing operation is the difference between:

$$\tfrac{1}{2}(L_1+L_2) \text{ and } \tfrac{1}{2}(L_3+L_4)$$

The effect of the expensive material and weld dressing compared with the cheaper material and no dressing is the difference between L_1 and L_4, and the difference between expensive material (no dressing) and cheaper material (with dressing) is the difference between L_3 and L_2. This is very efficient testing because all four results are being used to assess the benefit of the change of material and the same four results are also being used to assess the benefit of change in the finishing process. Using a classical scientific approach the test programme would have been as follows:

$$\left.\begin{matrix} A+ & B+ \\ A+ & B+ \\ A+ & B- \\ A+ & B- \end{matrix}\right\} \text{ to assess the effect of a change in } B$$

$$\left.\begin{matrix} A- & B+ \\ A- & B+ \\ A+ & B+ \\ A+ & B+ \end{matrix}\right\} \text{ to assess the effect of a change in } A$$

This means twice as many tests to obtain the same amount of information for choosing the material and finishing process.

This is all very simple and is termed *Factorial Testing*. It is much more powerful when trying to sort out three or more variables. Typically, a third variable in the example being considered would be the possibility of prestraining the component beyond the proof limit to investigate the benefits of residual stress, over and above that induced by the weld dressing operation. The components with prestressing are termed $C+$, and those without $C-$. The test plan then becomes:

	$A+$		$A-$	
	$B+$	$B-$	$B+$	$B-$
$C+$	L_5	L_6	L_7	L_8
$C-$	L_9	L_{10}	L_{11}	L_{12}

The results of the eight tests are L_5 to L_{12} inclusive and they are analysed as follows:

effect of:	*compare*:	*with*:
material (A)	$\frac{1}{4}(L_5+L_6+L_9+L_{10})$	$\frac{1}{4}(L_7+L_8+L_{11}+L_{12})$
weld finish (B)	$\frac{1}{4}(L_5+L_7+L_9+L_{11})$	$\frac{1}{4}(L_6+L_8+L_{10}+L_{12})$
prestrain (C)	$\frac{1}{4}(L_5+L_6+L_7+L_8)$	$\frac{1}{4}(L_9+L_{10}+L_{11}+L_{12})$
material and weld finish together (A and B)	$\frac{1}{2}(L_5+L_9)$	$\frac{1}{2}(L_8+L_{12})$
material and prestrain together (A and C)	$\frac{1}{2}(L_5+L_6)$	$\frac{1}{2}(L_{11}+L_{12})$
weld finish and prestrain together (B and C)	$\frac{1}{2}(L_5+L_7)$	$\frac{1}{2}(L_{10}+L_{12})$
material, weld finish, and prestrain (A, B, C)	L_5	L_{12}

This really does sort out the situation, but the penalty is that the number of tests is 2^n, where n is the number of variables.

with 2 variables, there are $2^2 = 4$ tests
with 3 variables, there are $2^3 = 8$ tests
with 4 variables, there are $2^4 = 16$ tests

On the other hand it is very efficient. In sorting out the individual effects of A, B and C, all 8 tests were used to reach a decision. In sorting out the effects of the pairs of variables, 4 tests were used to reach each decision. Only in considering the interaction of 3 variables were just 2 results used. In the classical scientific approach, only 2 results would be used to reach each decision, instead of 4 or 8 as shown above.

Reducing the number of tests

Often there may not be the time or money to contemplate 16 tests even though 4 variables are operative. This efficient and planned approach can still be adopted if other engineering considerations suggest that interactions between some of the variables will not occur. Suppose that engineering considerations suggest that $A+\ B+$ is different from $A-\ B-$ only because of the factor A alone, or B alone, and not due to any interaction between A and B. In the practical example we are considering this would not be reasonable because weld finish may have different effects on different materials. However, considering a test on a piece of rotating machinery, suppose:

A is the operating speed (rev/min) of the power unit
B is the environmental polution level
C is the bearing surface finish
D is the load torque

It would be reasonable to expect the effect of B to apply at all speeds A, and corrosion to proceed reasonably independently. Similarly it seems a reasonable guess that B and C will not interact and that A and C might or A or B and D might. The numbers of tests can now be reduced as follows (X signifying a deleted test).

		$A+$		$A-$	
		$B+$	$B-$	$B+$	$B-$
$C+$	$D+$	T_1	T_2	T_3	T_4X
	$D-$	T_5	T_6	T_7	T_8X
$C-$	$D+$	T_9	T_{10}	T_{11}	$T_{12}X$
	$D-$	$T_{13}X$	$T_{14}X$	$T_{15}X$	$T_{16}X$

Most effects can still be studied, even though the number of tests is reduced from 16 to 9. It is recommended that the keen reader completes the following table for agreement.

effect of:	*compare*:	*with*:
A	$\frac{1}{3}(T_1 + T_5 + T_9)$	$\frac{1}{3}(T_3 + T_7 + T_{11})$
B		
C		
D		
AB	assumed no effect	
AC	T_1	T_{11}
AD BC	assumed no effect	
BD		
CD		
ABC	not possible	
ABD		
BCD		
ACD		
$ABCD$	not possible	

The efficiency drops off with the number of tests because conclusions are being drawn from the difference between single tests. Nevertheless this programmed approach has enabled a number of effects to be studied which could not be done by a conventional method. The set of 9 tests has allowed the possibility to be investigated that:

(1) $A+$ $C+$ together produce a disproportionately large change compared with $A-$ $C-$
(2) $B+$ $D+$ together produce a disproportionately large change compared with $B-$ $D-$
(3) $A+$ $D+$ together produce a disproportionately large change compared with $A-$ $D-$

We have no time for factorial testing

This will be the common cry when the number of variables exceed 3 or 4. What can be done with 10 variables? For a full factorial test using only 2 values of each variable we need 2^{10} tests (1024). Suppose we start at the other end and say that we can afford time and money for 40 tests. In this case we proceed as follows:

test no	*variable*									
	A	B	C	D	E	F	G	H	J	K
1	+	+	−	−	+	+	+	−	+	−
2	−	+	−	+	−	+	+	−	+	−
3	−	−	+	+	+	−	+	−	−	+
4	+	−	+	+	−	−	−	−	+	+
.										
.										
.										
.										
.										
.										
40	+	+	−	−	+	−	+	−	+	+

The variables A to K can have two values each, designated + and −. Mark 20 counters +, and 20 −, and put them in a bag. Draw the counters out and mark down the *variable A* column + or − as indicated by successive counters. Repeat the exercise for columns B, C, D etc. In this manner a complete random selection of 40 test conditions has been established. The tests are run and produce the following reliabilities (y).

test no.	*reliability* (y)	A	B	C	D	E	F	G	H	J	K
						variable					
1	75	+	+	−	−	+	+	+	−	+	−
2	64	−	+	−	+	−	+	+	−	+	−
3	38	−	−	+	+	+	−	+	−	−	+
4	21	+	−	+	+	−	−	−	−	+	+
.											
.											
.											
.											
.											
40	63	+	+	−	−	+	−	+	−	+	+

The results (y) are plotted for each variable A to K. The values of reliability (y) are plotted for the first 4 variables (A, B, C, D) in Figures 5.4 to 5.7.

From these plots the following conclusions can be drawn:

(1) *A* has a modest affect and is only worth controlling if we are in a marginal situation (Figure 5.4)

(2) *B* can be relaxed and has no effect (Figure 5.5)

(3) *C* demonstrates an intereaction and must be further examined (Figure 5.6)

(4) *D* must be controlled precisely since the output is dominated by *D* (Figure 5.7)

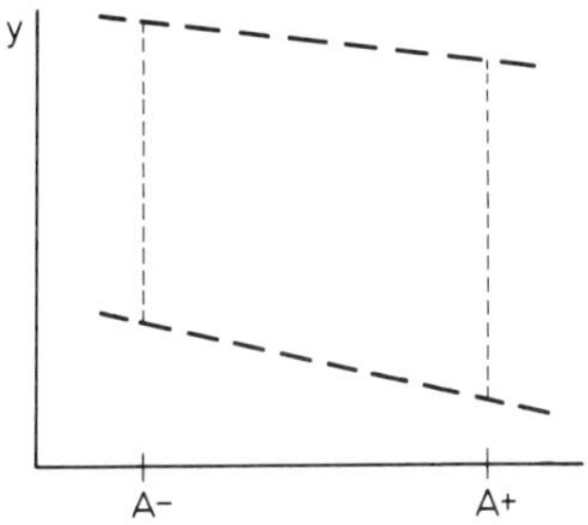

Figure 5.4 Parameter A has small influence on the test result

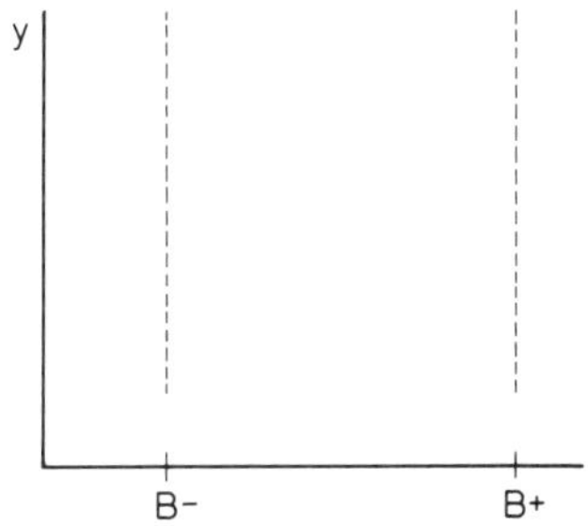

Figure 5.5 Parameter B has no influence on the test result

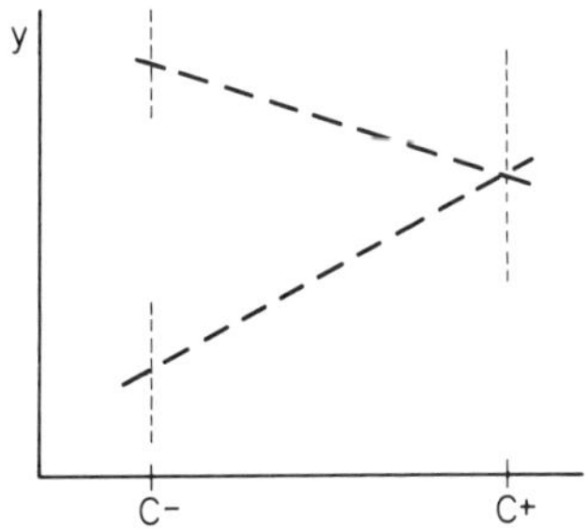

Figure 5.6 Parameter C has some influence on the test result and requires further investigation

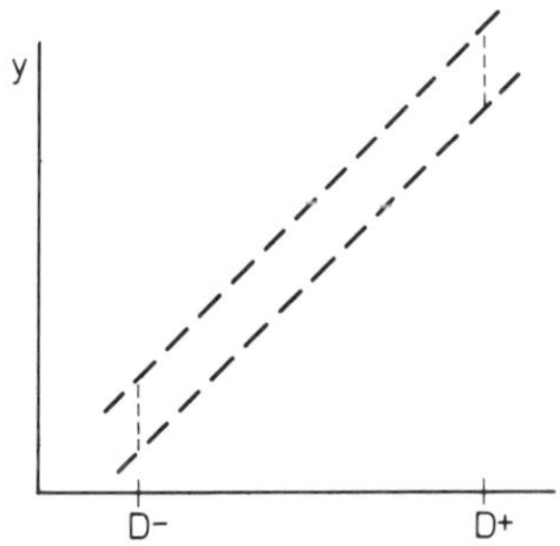

Figure 5.7 Parameter D has a strong influence on the test result

Even when a large number of variables may be affecting the reliability of the process or operation, if preplanned random tests are carried out in this way, the variables dominating the operation can be detected, and their effects quantitatively assessed. This is a very powerful tool which is well worth considering when the problem seems too great to start the test programme.

5.3 WHY DOES UNIT NO. 3 PERFORM BETTER THAN UNIT NO. 6?

Most development engineers frequently find themselves in this situation. If preplanning is adopted the answer can sometimes be made available from the build data. The performance or reliability (y) of a unit normally depends on a number of factors (x_i). x_i are the dimensions, clearances, interference fits, resistances, capacitances etc., that affect the unit performance and reliability. Without preplanning, x_i will in general not be known, except that they fall within the design tolerances. Suppose that the inspector records the values of x_i instead of merely declaring that the component is acceptable

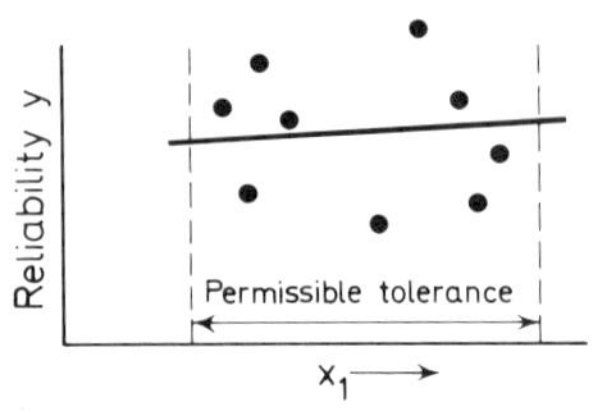

Figure 5.8 Reliability not very dependent on dimension x_1

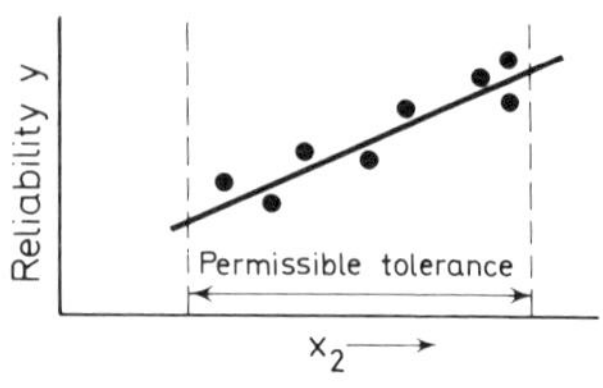

Figure 5.9 Reliability very dependent on dimension x_2

and within tolerances. The reliability (y) may or may not depend upon x_i (Figures 5.8 and 5.9). By drawing out these plots for each i (by computer if i is large), the largest value of M can be selected in:

$$y = Mx_i + C$$

In the example shown, the tolerances for x_1 can be relaxed but clearly the tolerance for x_2 must be decreased so that a suitably high reliability can be achieved for all units.

5.4 TAKE THE DECISION BEFORE THE TEST

'Have I really got an improvement, or is it just chance?' said the Development Engineer. When testing two units, to two slightly different standards B and C, which are almost identical, there is a 50% chance that B will look a little better than C (and of course, a 50% chance that C will look better than B). If the difference in performance between B and C is very large, then this difference is likely to be genuine, but this is not a typical case. Typically, the improvements being sought are of the same order as the scatter in results from the existing standard of equipment.

If the number of units tested is increased to two B units and one C unit then if there is no difference in real performance, the possible outcomes are:

better result	B	B	C
	B	C	B
worse result	C	B	B

which means the average B result is better than C one in 3 times (33% chance). If the test results show B to be always better than C, the chance that this is not true is 1 in 3 or 33%. In other words, if B is in fact no better than C, then 1 in 3 times when testing 2B and 1C the test results will suggest that B is better than C.

To check that you have understood this, try the following questions:

(1) What if we test 1B and 2C?
(2) What is we test 2B and 2C?
(3) What if we test 3B and 2C?

You may check your answers at the end of the book.

In this way, before the series of test starts, the decision can be made in the following manner: 'I will do five tests, 3B and 2C, and if all the B tests are above the C tests I will make the change from present standard C and accept the 1 in 10 risk that I am wrong. Otherwise I will not change from the present standard'.

Pre-deciding a course of action avoids the emotional trial of trying to decide what to do when presented with the test results. In particular, it avoids the frequent procrastination of trying just one more test which is the classical Development Engineer's way of avoiding a conclusion.

5.5 MINOR DEVELOPMENT PROBLEMS BECOME MAJOR SERVICE PROBLEMS

No matter what elegant techniques are used to sort out results and plan tests, there is the overriding consideration that the major unreliability problems should be attacked first. Once the really major unreliability problems have been solved, it is less easy to decide the priorities for action. Again, preplanning is needed if this data is to be available and the simplest system of reording is the best.

Problem	*occurred on*:				
sealing ring leak	2.10.69 3.2.69	1.2.70 4.3.70	10.9.70		
core plates cracking	1.5.70				
paint flaking	1.5.70	10.5.70	11.11.70		

The really major problems will be dealt with by development before the unit goes into service, but the minor development problems become the major in-service problems. Hence the necessity for the data on the minor development problems, since the usual state of

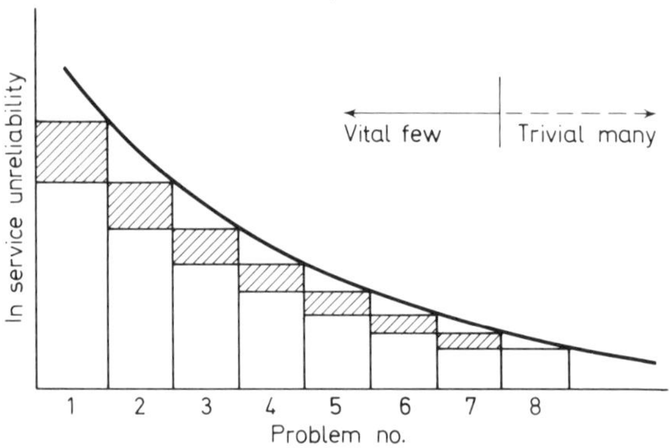

Figure 5.10 The vital few and the trivial many

affairs is shown in Figure 5.10. Real data for four families of gas turbine engines to show that this does occur is shown in Table 5.1.

Table 5.1 CONSTANT NO. OF PROBLEMS TO CURE HALF THE TROUBLE

Engine type	*first entered commercial service*	*no. of problems*	*% of overall basic unscheduled removal rate*
1967 *data*			
Avon	1958	7	50·7
Conway	1960	12	56·8
Dart	1953	10	51·4
Spey	1964	14	53·7
1968 *data*			
Avon	1958	7	52·8
Conway	1960	5	54·0
Dart	1953	11	53·2
Spey	1964	12	53·6
Total		78	426·2
Average		9·75	53·3
i.e.		9 problems account for 50% of the trouble	

In these highly technical products, with thousands of components being used all over the world, about 9 problems only cover half of all the unreliability on one of those families of engines at any one point in time. They are not, of course, necessarily the same 9 problems at a different point in time.

Without a record system to document problems and ensure that the follow-up action to cure those problems occurs in the right priorities, development can be a very inefficient procedure.

5.6 LOOKING FOR A DIFFERENCE

Any testing during which nothing goes wrong is confirmatory and will only reassure. Testing does not show the way to achieve improvement unless an attempt is made to search for some difference. Once a difference has been found and the cause of that difference has been established, then the path to improved performance or reliability can be identified. Differences in equipment reliability do not show up very quickly unless the equipment is very unreliable in the first place. To establish whether a motor car engine will last longer on a high or a low viscosity oil will take a long time. This sort of difference in reliability can only be satisfactorily established by testing in the field.

The sort of performance improvements which can be examined in the laboratory by test are those which can be defined as having a direct influence upon the reliability of the equipment. An example of this is the reliability improvement in a marine environment due to using the correct finishing process which renders the material resistant to salt laden moisture. Here the reliability of the complete equipment can be improved by simple laboratory tests on different finishing treatments. In this type of investigation the factorial system of testing can be applied to improve the efficiency of the test plan.

The amount of corrosion suffered by the material in a given space of time is likely to depend upon more than one factor. The production or process engineer will know from experience what vital few variables there are and may decide that they are:

(1) the cleaning process
(2) the type of paint
(3) the resistance of paint to minor damage

An efficient test programme for specimens of the material with different finishing treatment would be as in Table 5.2. The results C1 to C8 are the number of corrosion spots per inch after exposure

Table 5.2 TEST PROGRAMME FOR TWO PAINTS, TWO CLEANING STANDARDS AND TWO FINISHING TREATMENTS

	Trichloride clean		*Trichloride clean and sandblast*	
	Standard paint	*Heavy lead paint*	*Standard paint*	*Heavy lead paint*
Stoved	C_1	C_2	C_3	C_4
Unstoved	C_5	C_6	C_7	C_8

to salt spray. This form of testing allows the process engineer involved to choose the optimum finish for the longest and most reliable life, taking into account the possible interaction between the final cleaning operation, stoving and type of paint.

If there are no differences in performance on test, then it is not possible to draw any conclusions. An inferior result can be just as useful as superior result. A large drop in performance can tell an engineer more than a slight improvement and dedication to finding differences will shorten the problem solving time. If nothing goes wrong on a test then very little has been proved because an unreliable piece of equipment, or a piece of equipment manufactured in an unsatisfactory way may well survive for the length of time that can be spared for testing. One way of suitably motivating test engineers as they set off for the test rig, is to call out after them 'if you don't break something today you are fired'. There may possibly be a more subtle way of achieving the same end, but if no other approach brings the message home then the above may be adopted.

5.7 WILL THE TEST RESULT INFLUENCE THE DECISION?

The only reason for carrying out a test is to enable a decision to be made or under certain circumstances to confirm an existing decision. Before agreeing to a test programme it is economic to consider whether the results from the proposed series of tests will influence any decision that needs making. All too often the only action resulting from a test is the decision to do another test just to make sure.

Outside an academic environment there is little justification in doing any test if that test will not enable a decision to be reached. If this is not to the mental forefront when an engineer is deciding on a programme then the situation of just doing one more test, and then just one more, results in the information finally becoming available too late to influence the decision at all. This condition can be improved with conventional planning techniques by establishing the critical paths on a decision making network and establishing last dates by which the test results will be of any use. It all sounds terribly obvious and even childish, but if the next time you or one of your engineers plan to carry out a test, it may save money and valuable time if these simple questions are asked in a naïve way.

If the train will not arrive at the airport in time to catch the aeroplane, then do not waste time on the train journey and money on the train ticket.

5.8 TESTING WITH ARTIFICIALLY DAMAGED COMPONENTS

No matter what method of test planning is used there is an alternative to overstress tests which can be successfully used to detect, at an early stage, the problems which may lead to unreliability in service. This alternative is to use artificially damaged components. Sitting in a comfortable armchair skimming through this book, it may not sound a very revolutionary suggestion. But pause for a moment and think. What are your chances of successfully persuading your Development Engineer to knock his previous development units around a bit before he puts them on test? Not too good? In that case you need a few powerful arguments to help present your case.

There is often an excuse for failures that occur in service, such as *failure from a tool mark, poor surface finish, handling damage, slight misassembly* and so on. The reason for calling these excuses rather than reasons is that if an examination of the unfailed units shows that these minor faults are also present (as is usually the case) then the reliability of all the units is suspect. If a small amount of artificial damage had been introduced during the development testing, then early failures would have occurred during this phase and corrective action could have been taken before the units were put on the open market.

This is different from overstress testing, particularly where the failure mode is one of fatigue. The sensitivity of a component to fatigue failure may be heightened much more by local blemishes than by a uniform overstress. Fatigue is predominantly a surface effect, and small differences in the finish of components can make considerable differences in reliability. It could easily change the situation from 100% reliability to 0% reliability, which is a large change. Gas turbine compressor aerofoils for example, tend to fail from some local blemish on the odd occasion when a failure does occur. This blemish is usually caused by a small nick or damage mark caused by the ingestion of debris. However, from the very nature of take-off which involves tearing down the runway like an oversize vacuum cleaner, there are bound to be some instances of foreign body ingestion. It is assumed that the documented failures due to this cause on previous occasions will have caused the designer to choose a material which is not particularly notch sensitive from a fatigue point of view. The sensible thing to do therefore is to put a few nicks or blemishes on some of the blades on development units to see if the components are close to the threshold of failure. If the Development Engineer is not producing failures then it is time he looked for another job. That is one way of persuading him to try this approach.

A more technical way is to refer him to the concepts of French[15] who gives an explanation in simple statistical terms of the validity of this concept. He proposed that failure will occur or not occur depending on whether:

$$S = \frac{\sigma}{\sigma_L}$$

is greater than unity or not, where σ = alternating stress at point of failure and σ_L = fatigue strength at point of failure for life L. Since both σ and σ_L will be random variables, then S will also be a random variable and there will be a distribution of values (Figure 5.11). The failure frequency will be the area on the right hand side of the distribution. The probability of no failure in the first N units tested is $(1-\alpha)^N$. The distribution may be continuous as in Figure 5.11a, or non continuous where a single factor can produce a large change in stress level, such as may occur if a spacing washer is missed out on a few units (Figure 5.11b).

Suppose α is 0·1, then the odds are still slightly against a failure in the first six tested and even with this very unreliable unit no failure

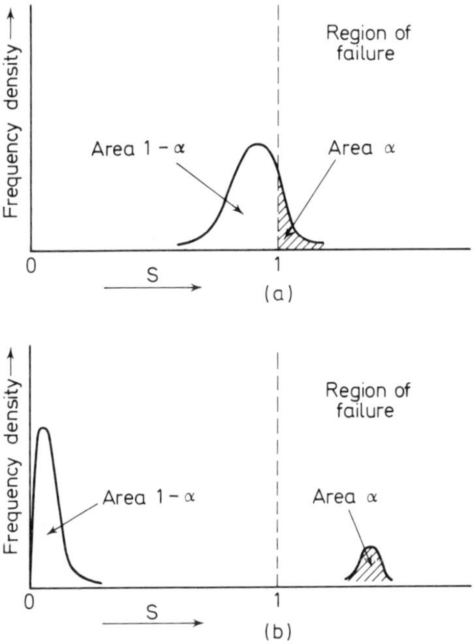

Figure 5.11 Hypothetical distribution of S, ratio of fatigue stress σ to fatigue strength σ_L

shows up in the early testing programme. Local treatment such as local roughening or etching may change σ_L enough to show the areas of unreliability during early testing. There will be no quantitative information about the unit reliability, only information that says something must be done to reduce the stress in a particular area of the component or unit. That is often sufficient.

6

What did you expect and what did you find?

6.1 HEADS OR TAILS

This is all about establishing the chances of being correct in accepting observed facts as true and not as just due to a strange quirk of fate.

The results of 10 tosses of a coin already discussed in Chapter 1 are:

toss No.	*heads*	*tails*
1	×	
2	×	
3	×	
4		×
5		×
6		×
7		×
8	×	
9		×
10		×

(a) Based only on these experimental facts, 4 times out of 10 (40%) the coin turns up heads, and 6 times out of 10 (60%) the coin turns up tails.

(b) Based on the first five tests, the findings could have been reversed, that is 60% heads and 40% tails.

(c) Based on the first three tests the findings were 100% heads and 0% tails.

The confidence in the conclusions drawn from 3 tests is clearly lower than from 5, and from 5 lower than from 10. We do not need to prove this, for both you and I have more than a sneaking suspicion that if only that test had done its job properly there would have been an equal number of heads and tails.

The results of a larger experiment involving 100 trials of 10 tosses each are as follows:

No. of heads in 10 tosses	*Results of 10 trials*										*No. of occurrences in 100 trials*	*No. of tails in 10 tosses*
0	0	0	0	0	0	0	0	0	0	0	0	10
1	0	0	0	0	0	1	0	0	0	0	1	9
2	1	0	0	0	0	0	1	0	0	1	3	8
3	0	2	1	4	0	3	2	1	1	0	14	7
4	4	4	2	2	5	4	4	1	2	4	32	6
5	3	1	1	2	3	2	0	5	4	3	24	5
6	1	3	3	1	1	0	3	2	2	1	17	4
7	1	0	2	0	1	0	0	1	0	1	6	3
8	0	0	1	1	0	0	0	0	0	0	2	2
9	0	0	0	0	0	0	0	0	1	0	1	1
10	0	0	0	0	0	0	0	0	0	0	0	0

From these experimental results the chances of a certain outcome can be assessed. For example:

(1) The chances of no more than 2 heads are, or the confidence in achieving at least 8 tails is: } about 4% or 1 in 25

(2) The chances of no more than 5 heads are, or the confidence in achieving at least 5 tails is: } about 74% or 3 in 4

(3) The chances of no more than 5 tails are, or the confidence in achieving at least 5 heads is: } about 50% or 1 in 2

Most engineers would feel pretty happy about accepting the result of a test which had been repeated a 100 times. However in this particular test, the results suggest that there are three chances out of four that at least five tails will appear, but that there is only one chance in two that at least five heads will appear. Now any reasonable considerations suggest that the chances of at least five tails in ten tosses are the same as the chances of at least five heads in ten tosses. Of if they are not then there is a good living to be made at your local friendly gambling club or casino.

If the test programme has not been carefully and statistically planned then it may or may not be reasonable to accept the blunt facts which the test results put forward. Equally well this suggests that to reach firm conclusions it is not particularly helpful to do a large number of tests since the wrong conclusion can still be reached.

This again is rather hard to believe—and is after a simple test like tossing a coin—what if we consider testing a complex piece of equipment? If the outcome of a test is going to be equally likely good or bad, then the coin tossing situation will apply. If a process only produces alternatively good and bad items, then this is a similar case.

Suppose that units are being passed off on a production line and on average 1 in 2 pass the final test, then the same numbers apply as in the coin tossing example to the chances of:

(1) at least 2 units in any consecutive 10 getting through the final test
(2) at least 5 units in any consecutive 10 getting through the final test

Hence the chances of being able to meet a particular deadline. This is single sided confidence, but it is often of concern that the outcome should lie between two limits, that is, the confidence that the number of heads will be greater than 2 and less than 8 is approximately 93%, since there is about a 4% chance of 0, 1 or 2 heads, and about a 3% chance of 8, 9 or 10 heads, based on the coin tossing results.

Fortunately in many engineering situations the chances of a good and bad outcome of a test are by no means equal, since the objective of a test is usually to show that a different way of doing something is better than the way used before. If there are 9 chances out of 10 that an improvement has been obtained anyway, then a test which gives an improved result is likely to be giving the correct conclusions. Similarly, if the improved result expected does not materialise from the test, it is likely that the test result is giving a true picture of the facts. The size of the improvement anticipated is of importance. If the improvement anticipated is only marginal, then it is less easy to accept the results at first sight.

The results from previous tests on a unit are shown in Figure 6.1, and the variation between units is of the order of 2% maximum to

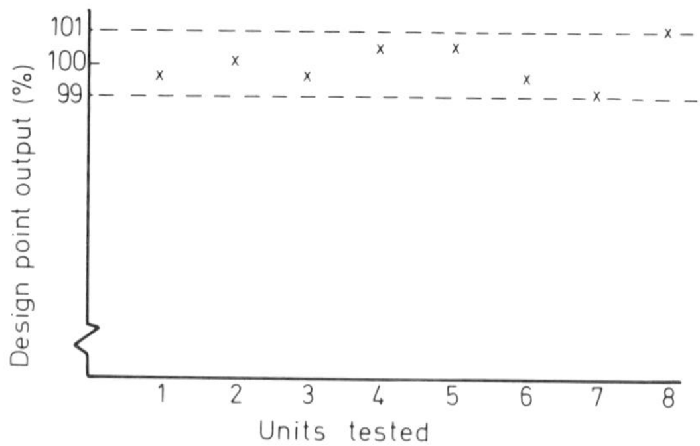

Figure 6.1 Variation in performance for the same standard of unit

minimum. Following an average unit, there is an evens chance that the next one will be better, and an evens chance that it will be worse. A unit which has an 'improvement' incorporated which is in fact no better than the original standard, has an evens chance of appearing better than average, and an evens chance of appearing worse, if the improvement anticipated is less than 1%.

However, the chances of two units with the modification fitted both appearing better when there is in fact no improvement is only 1 in 4, and there are 3 chances in 4 that the improvement is real. The improvement is worth having on this result because if on average three decisions out of four are good ones, then this is better than many engineers achieve normally.

6.2 A SIMPLE MATHEMATICAL APPROACH TO CONFIDENCE

At this stage it is worth looking very briefly at some simple mathematics to see what numerical juggling can be done easily and painlessly to assess if the results that have been produced mean very much in pure statistical terms. It is important to remember that where straightforward engineering judgement suggests a particular conclusion, then provided the statistical inference does not say that the conclusion reached is unlikely to be correct, then it is worth accepting the conclusion. Equally it must be remembered that where straightforward engineering judgement suggests that the answer could be either way, then do not make a change or spend money unless the statistical inference is that the results suggesting that the change is worthwhile are likely to be correct. Putting this more simply, do not be put off by your statistician being unable to commit himself or herself if your judgement tells you that you are correct. Also do not commit yourself to a line of action about which you are doubtful, unless your statistician has a high confidence in the numerical data produced.

High confidence means about 90% where you are very doubtful and about 80% where you are less doubtful. 90% confidence means that you are taking a one in ten gamble that you are wrong. I would suggest that if you are wrong only one in ten times then you are an exceptional decision maker, or else you are a bad Manager and never make any decisions until it is so obvious what to do that you are not needed to make decisions anyway. Perhaps you had better cut out this page before showing the chapter to your boss—or to your subordinates either come to think of it.

If simple statistics can be made to help increase the number of correct decisions that are made, and also increase the number of decisions that are made in a given period of time, then it is worth studying a little further.

Confidence limits can be assessed in many ways and your tame statistician can explain the best way in any particular application, but in this chapter only the *Binomial* and the *Poisson* approach will be examined. The Poisson approach is particularly easy to use for a quick answer which will be good enough to show whether it is worth spending any more time and effort on a more rigorous approach.

The Binomial distribution is described in many texts, and a summary of the distribution taken from reference 16 is; *if the probability p, of an event E is constant for all occurrences of its context event C, the outcome of which is either E or E then the probability of exactly xE's in n occurrences of C is given by the coefficient of t^x in the expansion of $(q+pt)^n$*. Clear? Well let's talk about it in terms of coin tossing. The probability of a head in one toss of a coin is 0·5. The probability of scoring five heads in ten tosses is therefore:

$$^{10}C_5 p^5 q^5$$

where $p = 0{\cdot}5$ and $q = (1-p) = 0{\cdot}5$, since q is the probability of not scoring a head in any one toss. Putting in the figures:

$$\frac{10\times 9\times 8\times 7\times 6}{1\times 2\times 3\times 4\times 5}(0{\cdot}5)^5(1-0{\cdot}5)^5 = 24{\cdot}5\%$$

The experimental result from the 100 trials reported before of 24% was very close to this theoretical value.

The Binomial expression can be used where the probability of an event is known with some precision, to establish the chances of a particular result from a series of trials or tests.

This is no good to us when n is very large, unknown or even tending to infinity. This is where the Poisson model can be of some help. Quoting the result only, again from reference 16; *the probability of exactly x occurrences of a statistically rare event in an interval of a stated length will then be the coefficient of t^x in the series*

$$\exp(-m)\,(1+\frac{mt}{1!}+\frac{m^2t^2}{2!}+\ldots)$$

and thus

$$p(x, m) = \frac{e^{-m}m^x}{x!}$$

where $np = m$ = finite number of occurrences × probability of occurrence if it were possible to determine it = Poisson mean if n tends to infinity. The main point to be noted from our practical view is that the probability of a small number of occurrences can be estimated. For example:

the probability of either 0 or 1 or 2 or ... $= \exp(-m)\left(1+\frac{m}{1!}+\frac{m^2}{2!}+\ldots\right)$

Confidence limits using binomial distribution

Taking first the case where no failures have occurred, we can write:

$$\text{probability of getting failure} = 1-\text{probability of no failures} = 1-R^N$$

where N = no. of trials; R = reliability per trial

The confidence we have in the reliability is equal to the probability of failure given that no failures have actually occurred. This is illustrated in Figure 6.2. As N increases R^N decreases and the

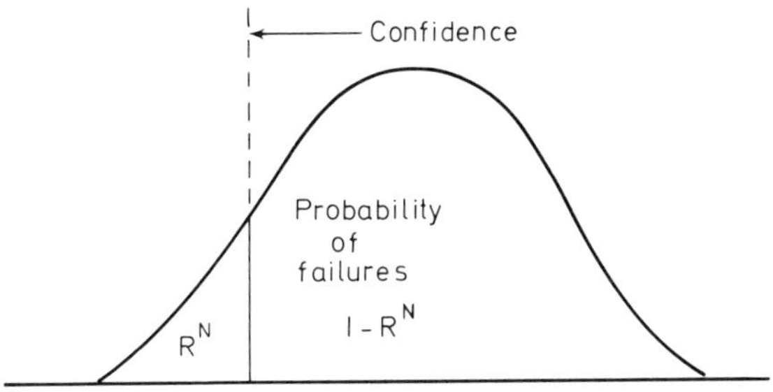

Figure 6.2 Confidence in the reliability is the probability of failure given that no failure has occurred

confidence level moves to the left in the figure increasing the confidence. We can accordingly write:

$$\text{confidence} = C = 1-R^N$$

We can set the level of confidence and work out a value for R with any given value of N. As an example, suppose we wish to know the reliability we have established, with 90% confidence, following the successful conclusion of twenty tests.

$$0{\cdot}9 = 1-R^{20}$$
$$R = \sqrt[20]{0{\cdot}1} = 0{\cdot}89$$

This is the sort of answer which makes a typical engineer wild. After no less than twenty tests, all that statistics can say is that he is 90%

confident that the reliability of the equipment that has not failed is not less than 89%—and we are paying him good money to reach that conclusion. But have patience and let us pursue this further to see what happens when failures do occur. How confident is anyone then that they can make quantitative statements about the level of reliability that has been achieved, if failures have been occurring on test? If we get one failure we can establish our confidence in the same way but now we must use the first two terms of binominal expansion and set the confidence equal to one or less failures. The equation now becomes:

$$C = 1-[R^N+NR^{(N-1)}(1-R)]$$

In the same way equations for the confidence in R can be written for any given number of failures. These equations can only be solved for R by iterative methods but as may be expected, they have been worked out and the results published in the form of charts and tables. For general reliability purposes charts are quite accurate enough and are more convenient to use than tables. Charts for 90% confidence limits single sided are given in Table 6.1.

This table shows that for 20 trials in which no failure occurred then one can be 90% confident that the reliability demonstrated is 0·8913 or 89·13%. This agrees with the value calculated from first principles. If one failure had occurred during these 20 tests, then the chart indicates that at the 90% confidence level the reliability demonstrated is 81·9% and for two failures is 75·74%. In all these cases the reliability demonstrated means reliability for a time period equal to the time that each unit was under test at the test load applied. If the test were an overload test then the reliability demonstrated would correspond to an appropriately longer life in actual service.

None of this sounds terribly useful in practice unless the number of units tested is large. When the number of units on test, or of more interest, when the number of units in service is large, then this approach can be useful. If the number of units is very large, then we move on to using the Poisson distribution.

Confidence limits for the Poisson distribution

We can use the terms of the Poisson instead of the binominal to calculate confidence limits. In this case we have the restriction that N must be a large number or that the number of failures is very small.

Given the number of failures we expect, NF, we can read the number that we may get with any degree of confidence, that is the confidence limit, directly from the Poisson probability chart shown

n	r = number of failures r = 0	1	2	3	4	5	6	7	8	9
1	0·1000									
2	0·3165	0·0513								
3	0·4644	0·1957	0·0345							
4	0·5626	0·3205	0·1426	0·0260						
5	0·6313	0·4158	0·2469	0·1124	0·0209					
6	0·6810	0·4892	0·3331	0·2011	0·0926	0·0174				
7	0·7194	0·5474	0·4039	0·2786	0·1695	0·1787	0·0149			
8	0·7498	0·5942	0·4619	0·3444	0·2395	0·1450	0·0686	0·0131		
9	0·7745	0·6319	0·5102	0·4011	0·3012	0·2100	0·1295	0·0608	0·0116	
10	0·7943	0·6627	0·5502	0·4487	0·3540	0·2621	0·1873	0·1159	0·0546	0·0105
11	0·8112	0·6897	0·5848	0·4890	0·4000	0·3171	0·2405	0·1695	0·1047	0·0495
12	0·8253	0·7124	0·6146	0·5245	0·4408	0·3611	0·2881	0·2189	0·1541	0·0955
13	0·8376	0·7326	0·6403	0·5556	0·4762	0·4007	0·3303	0·2639	0·2002	0·1418
14	0·8485	0·7497	0·6623	0·5826	0·5076	0·4355	0·3686	0·3043	0·2427	0·1852
15	0·8576	0·7642	0·6831	0·6073	0·5366	0·4673	0·4026	0·3413	0·2820	0·2256
16	0·8658	0·7780	0·7000	0·6286	0·5607	0·4959	0·4344	0·3744	0·3176	0·2632
17	0·8731	0·7897	0·7163	0·6481	0·5830	0·5208	0·4623	0·4058	0·3497	0·2974
18	0·8798	0·8004	0·7303	0·6661	0·6034	0·5435	0·4878	0·4331	0·3804	0·3285
19	0·8858	0·8101	0·7430	0·6814	0·6224	0·5645	0·5106	0·4587	0·4071	0·3584
20	0·8913	0·8190	0·7574	0·6956	0·6390	0·5841	0·5319	0·4815	0·4324	0·5846
25	0·9121	0·8531	0·8004	0·7516	0·7054	0·6596	0·6160	0·5734	0·5313	0·4911
30	0·9262	0·8765	0·8322	0·7906	0·7508	0·7114	0·6743	0·6381	0·6021	0·5673
40	0·9441	0·9060	0·8723	0·8404	0·8097	0·7794	0·7510	0·7232	0·6953	0·6685
50	0·9549	0·9245	0·8972	0·8713	0·8465	0·8215	0·7984	0·7757	0·7530	0·7311
75	0·9697	0·9491	0·9305	0·9129	0·8660	0·8790	0·8629	0·8473	0·8317	0·8168
100	0·9772	0·9617	0·9477	0·9344	0·9216	0·9086	0·8967	0·8848	0·8728	0·8613
150	0·9847	0·9743	0·9649	0·9559	0·9473	0·9386	0·9305	0·9226	0·9143	0·9065
200	0·9885	0·9807	0·9735	0·9669	0·9604	0·9538	0·9477	0·9416	0·9355	0·9295
300	0·9924	0·9871	0·9823	0·9779	0·9735	0·9691	0·9650	0·9610	0·9568	0·9528
400	0·9943	0·9903	0·9867	0·9834	0·9801	0·9768	0·9737	0·9706	0·9675	0·9645
500	0·9954	0·9922	0·9893	0·9867	0·9840	0·9814	0·9789	0·9764	0·9739	0·9715

n = number of trials

Table 6.1 LOWER 90% CONFIDENCE LIMITS OF ONE-SIDED INTERVALS

in Figure 6.3. If the reader does an exercise comparing the values he obtains from the Poisson with those given by the binominal he will find that the difference when $N = 100$ is small and that the values are almost identical when N reached 1000. The probability of x failures:

$$= \frac{e^{-m} m^{x}}{x!}$$

where m is the average number of failures. Where N is large and indeterminable the Poisson distribution can be useful for determining the confidence limits. Let us take as an example the case of some

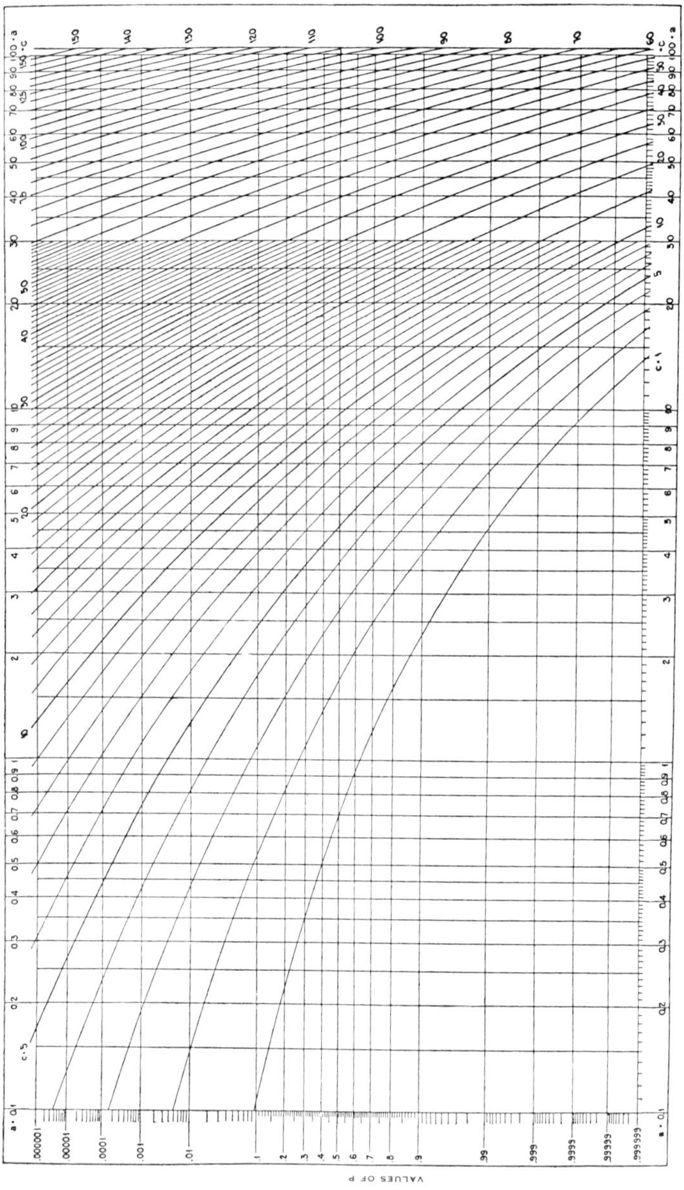

Figure 6.3 Poisson probability graph paper

equipment that has been giving trouble. We will assume that the trouble is not time related because, if this is so, we should have to take the age of the equipment into account. It takes a month to repair the equipment when it fails so we require sufficient standby sets to cater for the number of failures experienced in a month. We are prepared to take a 1 in 20 chance that we shall not have enough standby sets since these are expensive items to keep tied up as spares. The average number of failures per month over the last few months has been 15.

The problem can now be stated as if $NF = 15$, what is the single sided upper limit for 95% (1 in 20) confidence? From the Poisson probability chart we find that the lines for $NF = 15$ and $P = 0{\cdot}05$ intersect at between $C = 22$ and $C = 23$. Accordingly we require 23 standby sets if we are not to run a greater risk than 1 in 20 of not having spares available.

Exercises on Poisson chart

(1) 10 failures have occurred of 200 units to date. What are the chances of operator A, who has 20 units, having had at least 2 failures? What are the chances of operator B, who also has 20 units, having had 5 failures? Is it likely that operator B is different from operator A?
(2) An operator of 100 units has experienced 5 failures during 1969. What are the chances of at least 10 failures in 1970 if the true failure rate remains unchanged? Would you be surprised if he had 10 failures in 1970?
(3) I have a puncture of one of my car tyres every 5000 car miles. How many spare tyres do I need to take on a 500 mile journey for no more than 1 in 10 chances of running out of spare tyres? Is it reasonable to use Poisson?
(4) Repeat Question 3 from the beginning, starting with 'the chances of one failure are $500 \times 1/5000$ …'
(5) If I expect 1 failure, what are the chances of getting 0 failures? Now, how happy are you about standardising a modification or recommending a change just because no failure has occurred on test to date?

Answers

Just in case you have skipped the exercises, here is the answer to exercise 1 in some detail.
The average number of failures is 10/200 per unit. The expected number of failures for operator A is, on the average, $20 \times 10/200 = 1$,

that is, a = 1 and from the chart, the chances of having at least 2 failures ($C = 2$) are 0·27 or 27%. This does not sound very likely, but could easily happen by chance even if A is an 'average' operator.

Repeating for B, the chances of having at least 5 failures ($C = 5$) are 0·4%. This sounds very unlikely indeed.

It appears that operator A may be different from average and therefore merits a visit from the Service Engineer but that operator B is very different from average and his operation should be looked at very closely if the failures are serious ones.

The difference in this example is that looking at the numbers produces strong guidance about what course of action to pursue.

The Poisson approach is then the busy engineer's approach. Provided the sample is reasonably large, as is usual for an in-service case, and provided that the failures are not too strongly time or cycle related, then a couple of minutes with a Poisson chart can give a good first indication as to whether a collection of failures with one operator, or in one particular factory, or in one particular time period, are possibly due to chance, or possibly due to some local peculiarity which merits further investigation.

Once a real difference has been established between operators, or batches of units, or geographical location, or any other relevant factor, then it may be possible to locate the cause of that difference. It is then possible to take action and improve the reliability. The key to the problem is to locate a difference, and Poisson can be a real help to quickly assess whether a difference is real or apparent.

7

Maintainability and Reliability interaction

7.1 CONFLICTING IDEAS

'Maintain your equipment frequently and it will not let you down'. 'Do not play about with your equipment—you will only upset something'. These are opposed views on the influence of maintenance on the reliability of equipment and both can be correct for (different) specific circumstances.

If I maintain my car exactly as recommended by the manufacturer, then every three thousand miles I remove the front wheels, grease the nipples, remove the brake drum and blow out the dust, check the tyre for cuts, stones, glass, correct inflation pressure and then replace the drum and wheel. This will produce a high reliability of the steering, suspension, brakes and tyres. Unfortunately it produces a low reliability in keeping the wheel attached to the axle for the handbook omits to point out that after only a few removals and replacements, the threads on the wheel retaining nuts strip off. If no further action were taken other than the specified maintenance, then there would be a serious reliability problem which could have unpleasant consequences. Thus the reliability of some items is improved by maintenance and the reliability of others is degraded by maintenance.

Another car example is the unhealthy interest some owners have for removing the sparking plugs at frequent intervals. This is very nice for the plugs but not so good for the cylinder into which particles of grit will fall when the plug is removed. Turning now to the factory, a similar situation may exist where there may be more unreliability introduced by the addition of grit, grinding particles, turnings, chippings and so on into the machinery, than is ever cured by cleaning or oiling. Another example is the owner of some ducting who was for ever opening and closing the maintenance door to see if everything was all right, with the result that the door seal failed. There would have been no problem if the owner had not insisted on over maintaining his equipment.

Routine planned maintenance may stop the productive effort from a particular machine tool or production unit. The total down-time for planned and scheduled maintenance may be in excess of the down-time associated with the odd failure due to zero or minimum maintenance.

In any event, it may be that the failures are not associated with too much maintenance or lack of maintenance, but rather with the progressive deterioration which could have been monitored. If the monitoring indicates that a failure is imminent, then preventive action can be planned to avoid the deterioration progressing to the stage of complete failure.

7.2 ON CONDITION MAINTENANCE

This is the thinking which led up to the concept of *on condition* operation, associated with condition monitoring. A number of papers have been written by enthusiastic maintenance engineers who have successfully adopted this approach. One of the early ones was the TARAN programme[20]. Test And Replace As Necessary was introduced by United Airlines as a way of reducing the cost and time spent on the overhauling of hydraulic equipment. In addition, it reviewed the failures associated with *finger troubles.* Finger troubles are those producing unreliability in the type of failure distribution shown in Figure 7.1. Before the introduction of the TARAN programme much equipment was being dissembled and reassembled for inspection irrespective of whether it was working satisfactorily or not.

The parameters that can be used as an indicator of failure depend upon the type of plant or equipment being used, but a selection are:

Vibration—measured by vibration transducers attached to the machinery.
Temperature—of oil, air, gas, electrical windings.
Oil—consumption for leaks or wear: spectrographic analysis for signs of bearing failure or excessive wear.
Noise—by ear or by frequency band analysis.
Performance—fuel or power consumption: pressure rise or fall: temperature rise or fall or power output.
Reaction time—time for an actuator to move: time for a rotating device to reduce speed: time for voltage decay.

Visual—looking, borescope, intrascope (flexible device which can be inserted into machinery and has a small light at the problem end and transmits the image to an eyepiece outside the equipment).

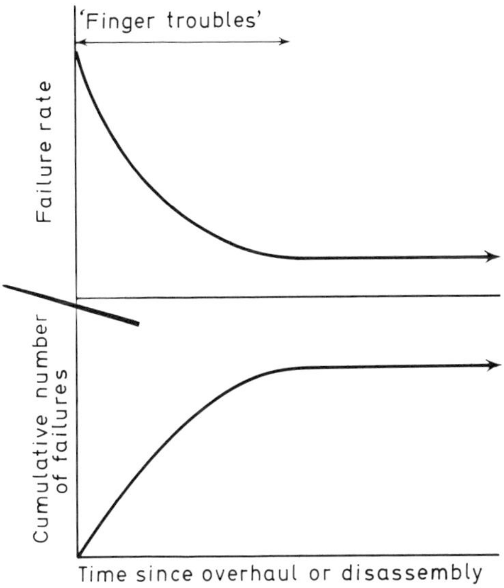

Figure 7.1 Failure rates associated with 'finger troubles'

The snag is that the fingerprint or monitored pattern associated with a particular mode of failure can only be established by monitoring the parameters before a failure occurs. Once it has occurred attempts can be made to associate the measured fingerprint with the appropriate failure mode. Until then there may be a number of false alarms when observed patterns abnormally deviate before any obvious signs of failure are present—perhaps for reasons not associated with impending failure.

The qualitative ideas on maintenance and reliability interaction are:

(1) Minimum possible routine maintenance to maximise the time that equipment is available for production or use and to minimise the *finger trouble* unreliability.
(2) Maximum use of any parameter which can be used to give warning of impending failures in preference to routine inspection or overhaul.

7.3 QUANTITATIVE TRADE-OFFS OF *R* AND *M*

Let us now consider the case where maintenance can restore equipment or machinery to the *as new* condition without complicating the issue by the introduction of finger troubles. It is rather difficult to quantify reliability in this case without also specifying the maintenance interval. This is best illustrated by an example. Failure Rate will in general increase with age (time since new) as shown in Figure 7.2. If maintenance is carried out at 10 000 hour intervals,

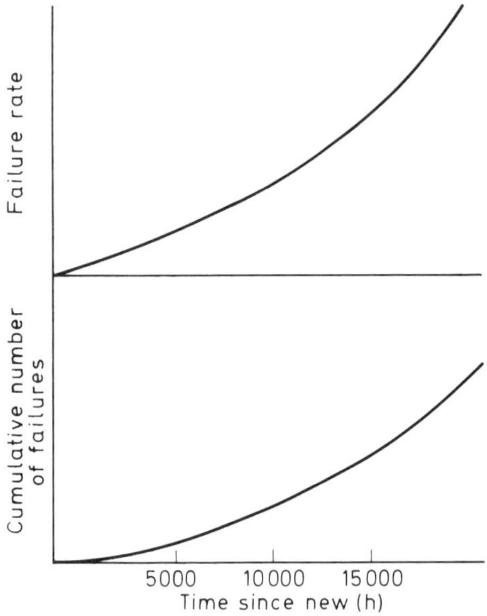

Figure 7.2 Failure distribution and rates if no maintenance

then the failure rates and number of failures will be as shown in Figure 7.3. If the maintenance interval is reduced again to 5000 hours, then the corresponding failure rates and cumulative number of failures will be as in Figure 7.4.

The average failure rate is vastly different for no maintenance, overhaul at intervals of 10 000 hours, and overhaul at intervals of 5000 hours. What is the true reliability of the equipment? There is in fact no such thing in this case. The situation is that for a known correlation of failure rate and time since overhaul, the reliability can be set at the level acceptable to a particular form of operation by adjusting the maintenance interval. If no hazard is involved, and a

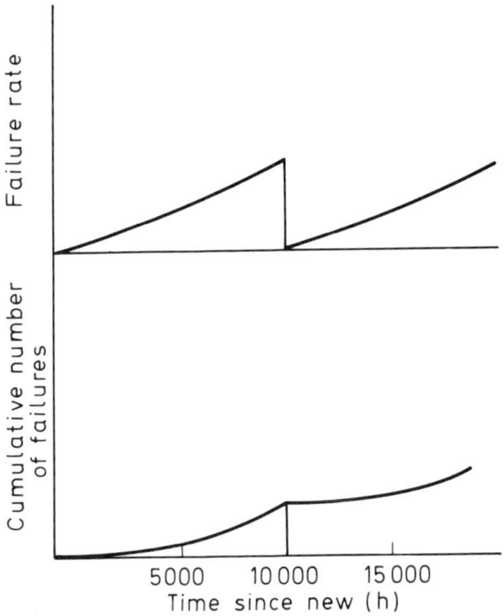

Figure 7.3 Failure distribution and rates if maintenance at intervals of 10 000 hours restores to the as new condition

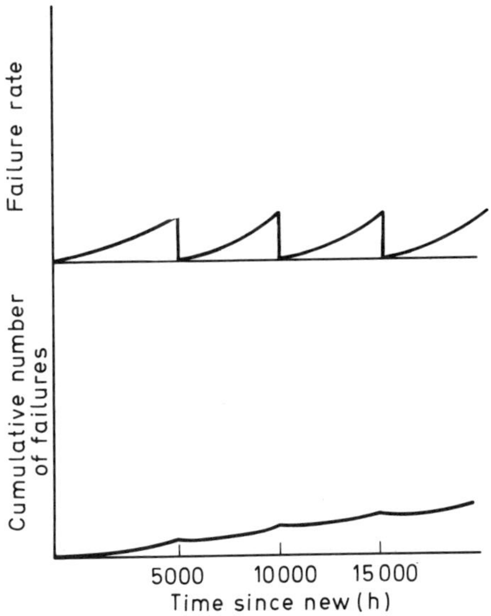

Figure 7.4 Failure distribution and rates if maintenance at intervals 5000 hours restores to the as new condition

failure can be tolerated, then the correct maintenance interval is that which produces the minimum operational cost (Figure 7.5).

If maintenance labour costs are of paramount importance, then the possibility of achieving 100% reliability by means of standby

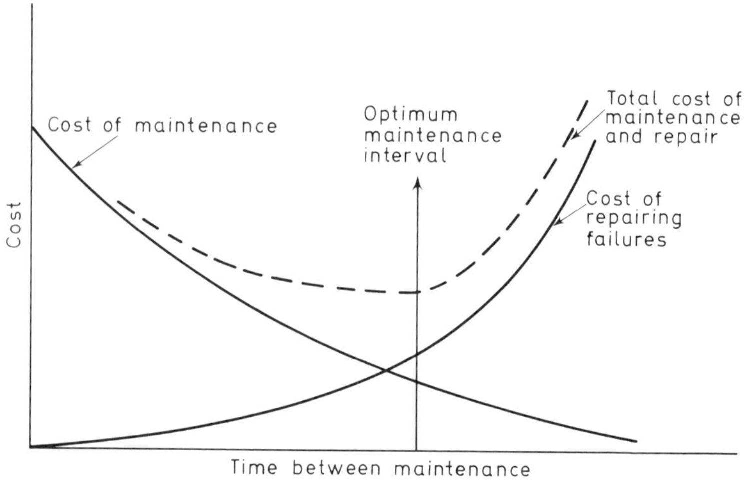

Figure 7.5 Optimum maintenance interval can minimise total cost of maintenance and repair

redundancy may become an attractive financial proposition. This is equally true where the cost of having plant out of commission is very high, and a small amount of down-time would cost more than the true cost of either keeping passive redundant standby equipment idle for much of the time, or running active redundant equipment at part load for much of the time.

In practice, the operator of expensive equipment wants to have his cake and eat it, all without the bother of even cutting it. Huddie[21] shows how the reliability of equipment can increase with time since new, and at the same time the maintenance interval can increase (Figure 7.6).

This is contrary to any reliability and maintainability theory and implies that a continuous development activity is taking place. The curves relate not to one standard of equipment, but to a continuously changing standard. The data after 3 or 4 years will in fact refer to a variety of standards of equipment, some of the units being built to a much later modification standard than others. This will be due to a combination of deliveries of the equipment over a period of time and to the operators' staggered modification incorporation programme.

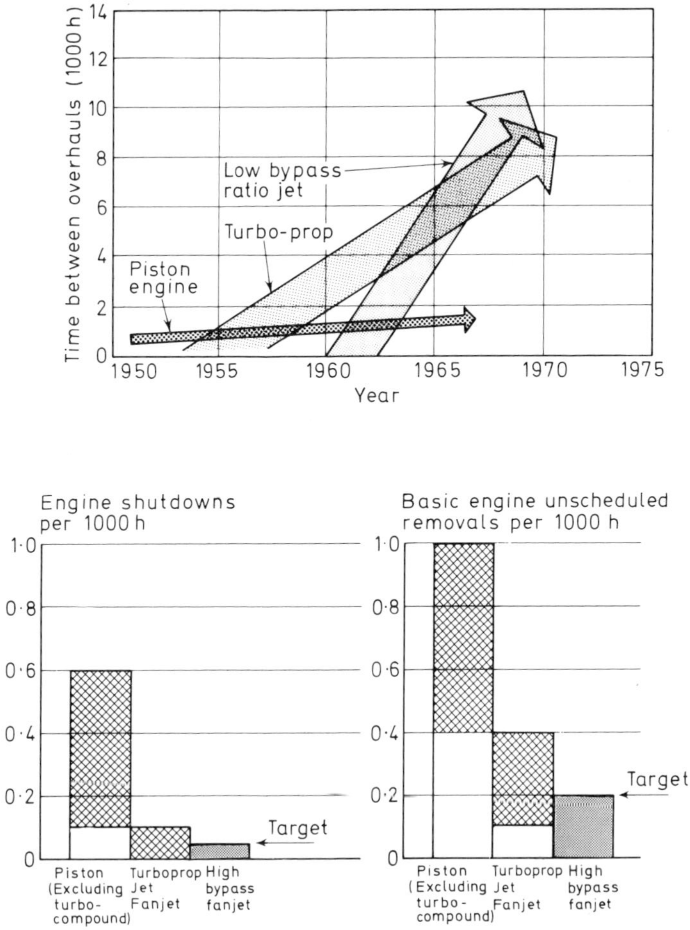

Figure 7.6 Maintenance intervals can increase at the same time as reliability increases. Data shown is for aero engines, showing the dual effect of progressive technical improvement

7.4 THEORETICAL RELATIONSHIPS BETWEEN *R* AND *M*

Failure rate $f_1 = F(t)$

where t = time since new

Failure rate $f_2 = F(t - nt_m)$

where t_m = maintenance interval

n = number of maintenance operations carried out

This assumes that maintenance is carried out which restores the equipment to the as new condition.

Reliability $= R = 1 -$ unreliability, and for equipment being maintained at intervals:

$$= 1 - \frac{1}{N}\int_0^T f_2 \, dt$$

$$= 1 - \frac{1}{N}\int_0^T F(t - nt_m)\, dt$$

where T = total time of operation of each unit
and N = number of units

Consider now a linear relationship between failure rate and time since last overhaul:

$$F(t) = kt$$

$$F(t - nt_m) = k(t - nt_m) \text{ (Figure 7.7)}$$

$$\text{Reliability} = 1 - \frac{1}{N}\int_0^T k(t - nt_m)\, dt \text{ where } T = nt_m$$

$$= 1 - \frac{k}{2N} nt_m^2$$

which shows that as maintenance interval t_m tends to 0, Reliability tends to 1 and as t_m tends to ∞, Reliability tends to 0

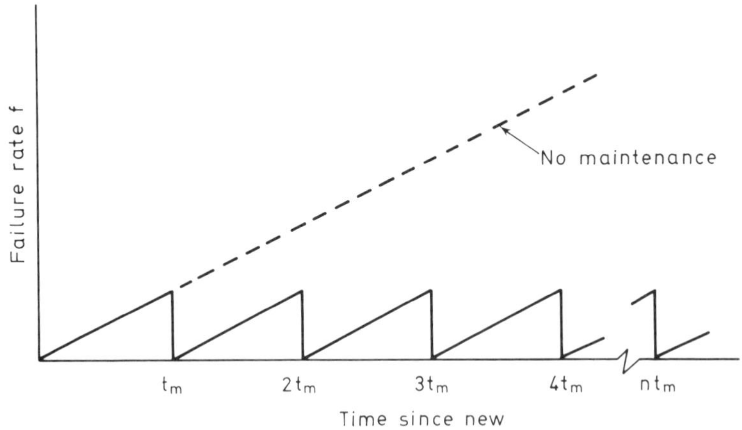

Figure 7.7 The effect of overhaul time on linear failure rate curve

Similarly, expressions can be derived for other relationships between failure rate and time.

This merely shows quantitatively what has been stated earlier in a qualitative fashion, namely that with a short maintenance interval, the reliability tends to 1 (100%), and for a long maintenance interval the reliability tends to zero (all units failed). If the relationship between failure rate and time since new is known, or can be predicted, then it is worthwhile studying the correlation between reliability and maintenance intervals. If no correlation can be established, then there is a real possibility of curtailing the amount of maintenance carried out. Equally, if a strong correlation exists, there may be a good case for more maintenance in order to have less failures.

As in most other concepts of reliability, if quantitative information is obtained then the best decision relevant to the prevailing circumstances can be made. To make decisions you need numbers.

7.5 THE FLEET OPERATORS VIEW OF MAINTENANCE

The main conceptual advances in recent times on maintenance and maintainability programmes have come from the Airlines of the world where maintenance forms a substantial portion of their operating costs. They have more data available than is typical in many other industries and so can reach more numerate decisions and rely less on preconceived ideas and hunches. The TARAN programme has already been referred to and Davis[20] states: 'one of the first things we realised was that we were not finding much to complain about in the performance of most assemblies before we took them off of airplanes at overhaul. When we took them apart, we didn't find much wrong there either, even though there were usually some pieces which didn't meet new part limits any more. So we replaced the pieces, put the components back together and tested it. Sometimes it worked fine—other times it didn't work as well as it had before we fixed it. Not only that, we already knew as did a lot of other people, that when we disturb the plumbing in the airplane to make the component change, we were exposing ourselves to new troubles there too. We began to think that there must be a better way. We began to look for solid reasons why we were arbitrarily removing components that seemed to be normal, and replacing them with components that might turn out not to be normal (even though we had tested them). And in the process we were increasing our chances of trouble in the immediate neighbourhood of the change. If nothing else, we were fairly certain that the cost of overhauling these systems was higher than it needed to be'.

At about the same time Wallach,[22] also of United Airlines was explaining that if failures are not time related then overhauling is a pretty useless pastime. He explains: 'historically, scheduled maintenance was accepted as the most important method used to accomplish our objectives. However, experience has shown this line of thought to be false. In most cases the reliabilities of complex components and systems, including overall measures of reliability of the complete airplane either show no deterioration with increasing age or very slow rates of deterioration. Complex components seldom show any pronounced adverse relationship between reliability and age'.

The availability of computers has enabled maintenance planners to study the failure patterns of the equipment they are maintaining. Robertson and Olson[23] have produced some startling figures relating to the correlation of failure rate and age. 'To date we have analysed the computer output for about 37 selected high cost units. The results showed that 95% of the units involved had either a constant or decreasing failure rate as time increased. For these units it is difficult to justify a specification that proposes to remove the unit from the aircraft for overhaul at some arbitrary time limit'.

In the following three years events moved rapidly and not only the airlines, but also the airworthiness authorities became convinced that a cheaper and more reliable operation could be achieved by reducing the amount of routine overhauling being done to their equipment, Edwards[24] of the Air Registration Board had spoken strongly in 1967 against unnecessary disassembly of equipment: 'The processes of physical inspection can entail considerable disassembly which almost inevitably causes damage and reduces the quality of a part. Disassembly for inspection is a time consuming and costly activity which in itself, contributes nothing to airworthiness. It follows that physical inspection involving disassembly should be performed no more frequently than is absolutely necessary and that provision should be made for inspection, where possible, without disassembly'.

In the space of only a few years the main basis of maintenance procedures has been changed with the full approval of the authorities concerned. It will be interesting to see if other industries which spend large sums of money on maintenance will follow the path that the airlines have set. The armed forces who traditionally maintain and maintain and maintain their equipment are being quick to follow and considerable economies will be achieved there. Consultants in the process industries using the simple concepts outlined here are going to make a lot of money for themselves whilst saving it for their customers.

The main requirement is an enquiring mind and the determination to collect facts about the failure patterns that arise in a particular plant and a firm disbelief that all maintenance necessarily results in a more reliable and inexpensive operation.

The whole of this chapter can be summarised as follows. If the failures are time related—maintenance can improve reliability. If the failures are not time related—maintenance may reduce reliability.

8

Analysis of service data

8.1 FEEDBACK AND FEEDFORWARD FROM SERVICE EXPERIENCE

It is vitally important to know how the product is surviving in service. The dissatisfied customer may well say little but just buy elsewhere in future. No news may be good news, but it may not. In the hurly-burly of industry it is sometimes simpler just to put the inadequate equipment on one side and leave it to rot until the space is required for something else. Equally, very few of us write to a supplier to say that his product is giving excellent service. Is no news good news or bad news? Without information it is not possible to be a rational manager and to make correct decisions.

The most usual ways of retrieving information from the customer, or about the customer's operation, are:

(1) routine reports from the customer
(2) routine reports from the supplier's service engineers
(3) audit by an audit team
(4) spares orders

Routine reporting from the customer is certainly the cheapest for the supplier, but is not without its problems. The information received is only that which the operator chooses to send, and furthermore it cannot be confirmed by the supplier. The diagnosis by the customer and operator may be incorrect, and 50% correct diagnoses is perhaps the figure to expect. This is often good enough to make the decision on where to put the engineering effort to rectify the worst problems. The biggest advantage of using customer data is that the customer reports on those aspects which concern him most, which are usually those that are costing him money. These problems may not be technically fascinating, but they are the ones to which particular attention should be paid by the supplier. Another advantage is that by using the customer's data there is less possibility of a dispute arising as to what happened and how many times it occurred, since there is a single source of information. Data from more than one source can result in much time consuming effort being devoted to cross checking two different sources of information. Routine reporting from the supplier's service engineer is reasonably reliable, in that the proportion of correct diagnoses should be higher than where the customer makes his own diagnoses. There is still the problem of different service engineers reporting to different yardsticks unless effort is made to achieve a standard against which to report. This can soon become an expensive business which may not repay itself. Reporting is a chore which many service engineers would willingly avoid, and after a long day putting a problem right, it has to be a highly motivated engineer who will then report in detail the results of his day's endeavours.

In order to maintain the standard of reporting it is vital to give the engineers something back for their efforts so that they see some benefit arising out of all the paperwork which routine reporting entails. This means that the system should give them back the experience reported by other service engineers to say what problems they should be looking for in the near future. Routine reporting by service engineers is nevertheless a very acceptable method of retrieving information.

Audit teams sent in for a specific period of time will normally produce the highest standard of information. The members of the audit team know that they are there for a limited time period and do not have to divide their efforts between fixing the problem and reporting on it in the context of how the total operation is going. They are professional fact finders. The snag is that they are only around with a particular customer for a limited period and the assumption is made that the events recorded during that period are typical of any other period. The limited period visit may result in atypical findings being reported, but if the audit procedure is routine at discrete intervals, the chances of being misled are reduced. The chances of being misled are reduced further if the time of the visit is selected at random. This can be done by drawing names and dates from a hat, or using random numbers tables if you wish to look knowledgeable about statistics. If no other source of information is available, then spares orders can be used to give some idea of what is going on. If spares sales of a few particular components are above the norm, then they can act as an indicator and make it worth sending the service engineer to see what is going on. Spares orders alone are unlikely to give much warning that serious problems are developing since by the time the spares are ordered the trouble has set in and the customer is screaming and reading the small print in the contract.

Since many operators of capital equipment now maintain minimum inventories of stores, there is little breathing space in the system to allow time for corrective action to be taken. His small stock of spare parts will soon be used up if the problem is repetitive. The problem is most acute with large expensive components, for nobody likes to have many of these lying idle, so if a problem does strike with such a component which rapidly appears in all other components, the spares supply situation may become impossible. This is the main danger in relying on spares orders alone to provide a story of how equipment is behaving.

The worst thing of all—and a common occurrence—is to collect information and then do nothing with it. Information can be expensive stuff to collect, equally expensive to process, and so before collecting any, it is important to decide why the information is to be collected, whom it is meant to inform and how that information is going to reach the correct destination in a timely and inexpensive manner. The sins of collecting no information and collecting information but doing nothing with it seem to be equal. If the information collected never contributes to the action, then spend no more money on collecting that information.

The Service Manager needs information to plan and control his product support operation effectively and efficiently, and he needs adequate data to ensure that engineering or other action proceeds along lines destined to cure the main customer dissatisfaction.

In order to manage an operation, service data is analysed at two levels:

(1) At a simple level to indicate those areas in which problems lie, and those areas which are relatively problem free.
(2) At a more detailed level to help establish the causes of failure, and to predict what failure pattern will develop if no action is taken.

8.2 PROBLEM INDICATION

Management need to know how an operation is going. If all is well, that is all anyone wants to know. If all is not well, the problem areas need to be highlighted. The fixing of *alert levels* is one of the simplest approaches (Figure 8.1). The trick is to set the alert level close enough to the normal level so that an alarm can be raised in a timely fashion, but not to set it too close to the acceptable so that almost every month the situation is reported as unsatisfactory (Figure 8.2).

An improvement on this is to maintain a *trend analysis*, and to raise warnings on changes rather than on absolute levels (Figure 8.3). This can be more sensitive and can provide an early warning of

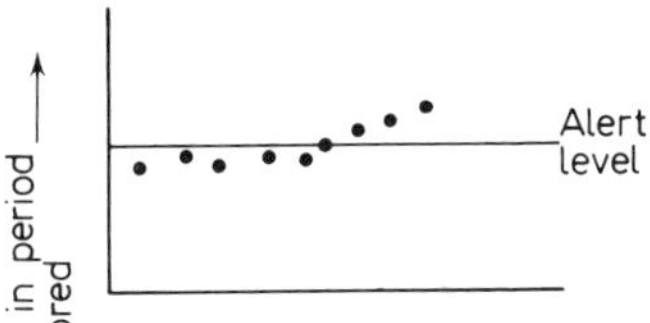

Figure 8.1 The fixing of an alert level allows routine monitoring to signal a problem

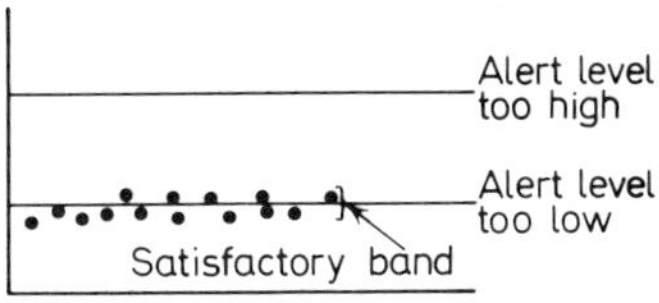

Figure 8.2 Alert levels are usless if set too high, or too low

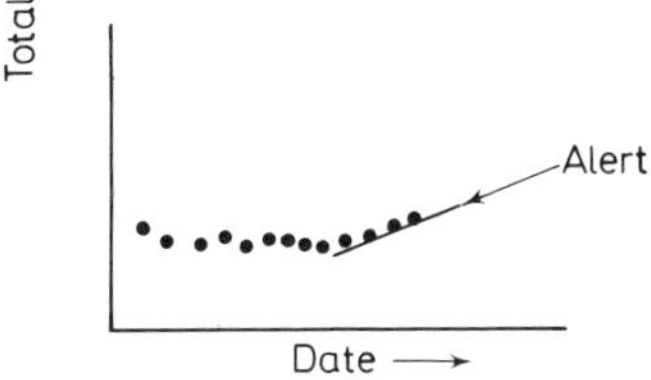

Figure 8.3 Trend analysis need not be related to absolute levels of reliability

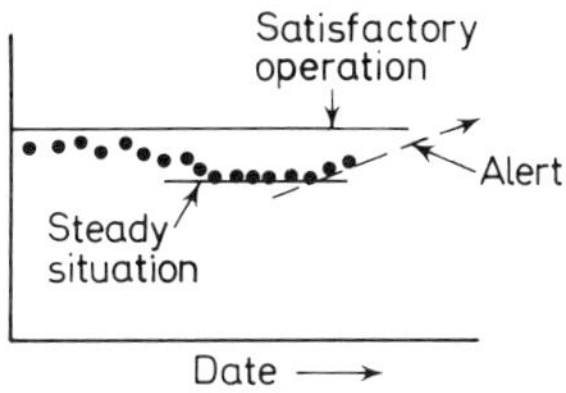

Figure 8.4 Trend analysis shows the necessity to take action even before the reliability deteriorates to a satisfactory level

an adverse trend even when the absolute level is better than required (Figure 8.4).

Confidence can be added in complex ways or in the following simple manner. If the operation is at a steady level of reliability with only small changes up and down due to chance:

There is a 1 in 2 chance that the next check will appear worse than the present.
There is a 1 in 4 chance that both the next two checks will appear worse.
There is a 1 in 8 chance that the next three checks will appear worse.

In other words, in Figure 8.4 we are only taking a 1 in 4 chance of being wrong in saying 'things are getting worse'.

8.3 IN-SERVICE RELIABILITY ASSESSMENT

The chief reason for assessing in-service Reliability is to help Management decide what needs to be done to improve the reliability of the product to an acceptable level, or to maintain an achieved acceptable level. This being so, it is worth following through the ideas of Brook[12] and spending a little time thinking about the *Apparent Reported Reliability* of a product.

The main point to grasp when looking at the reliability of an existing in-service operation is an appreciation of the variety of modification standards that may be contributing to the overall assessment of reliability. In other words one must be careful to separate what is being seen in practice, and what is theoretically obtainable if the customer is prepared to buy and fit all the modifications on offer. Usually he is not prepared to fit modifications as soon as they are standardised and made available because of the cost of the disruption to his operation. As far as gas turbines are concerned, the customer usually chooses to raise the modification standard of the engine at the time that it comes to the maintenance shop for one reason or another, not necessarily related to the component he wishes to modify. The tendency is therefore always to *over engineer* a product in so far as an engine built to today's standard, incorporating all available modifications, would be considerably more reliable than the partly modified or unmodified engines which are in service. If the engineering action is programmed on apparent reliability levels, there is a tendency to produce more engineering fixes than are actually going to be used. This is expensive and wasteful. Measured differences between actual reported in-service reliability levels of gas turbine engines and those levels which could be achieved if customers fitted all the modifications on offer are shown in Table 8.1.

Table 8.1 APPARENT AND THEORETICALLY OBTAINABLE RELIABILITY LEVELS

Engine	*Reported basic unscheduled removal rate in* 1968 *(per* 1000 *h)*	*Basic unscheduled removal rate in* 1968 *for problems without an adequate fix (per* 1000 *h)*
	(apparent level)	*(theoretically obtainable level)*
Avon	0·307	0·136
Conway	0·195	0·100
Dart	0·150	0·056
Spey	0·305	0·163

The third column in Table 8.1 is therefore the unscheduled engine removal rate in December 1968 for an engine built to December 1968 standards. It is interesting to note that based on the experience shown in Table 8.1 the highest modification standard of an engine in any one family could have an unscheduled removal rate of between 37% and 53% of the actual apparent removal rate. In other words, the theoretically obtainable rate is between only 37% and 53% of the apparent removal rate.

It is important to discount the reported problems for which a cure is available when assessing the engineering work that remains to be done. This presentation of information helps the customer to decide whether or not he is prepared to buy the latest modification by presenting him with clear alternatives—a higher reliability if the modifications are bought and fitted, and a lower reliability if he prefers to spend no further money on modifications.

'Look at the unreliability—we've got to spend some more money to help support our customers' is only tenable if unreliability is expanded to mean unreliability due to problems for which we have offered no cure. It sounds obvious but is not always so obvious to the engineering manager brought up to think more of technical problems than of obtaining the maximum value for each pound he spends of his employer's money. It is not a question of reliability at any price, but rather of reliability at a price which the supplier and customer can both afford. The market is only entitled to the standard of reliability that it is prepared to pay for, so it is important to ensure that the market understands the theoretically obtainable reliability that can be offered at a particular stage of a product's life cycle. This information prevents misunderstandings arising from apparent reliability which is the reliability that is more usually calculated and presented.

8.4 DETAILED ANALYSIS

As discussed earlier, the aim of detailed analysis of service data is to establish a credible failure pattern to explain the failures that have occurred. This pattern must satisfy:

(1) engineering considerations
(2) the numerical facts
in that order of importance.

When (1) and (2) are both satisfied then there is a good chance that the correct degrading parameter causing the failures has been located. Any extrapolation of this pattern is likely to give a more accurate forecast than intuition or blind guesses (both are frequently used forecasting techniques).

The aim of the analysis must be primarily to see if the true failure rate is: (1) decreasing; (2) steady or (3) increasing. This in itself may be enough for a correct management decision to be taken, and the absolute values of unreliability may not be required to a high degree of accuracy to influence the decision making process.

Some of the ways of looking at the data are considered in the following example,[25] which refers to a problem that was causing failures in 1967 and 1968. This was a steel component that was failing due to fatigue, and the data for the first 7 failures were as follows:

Table 8.1 FAILURE RECORD

Failure No.	*Date of failure*	*No. of hours run since new*	*No. of hours run since last build*
1	12/68	4590	2510
2	8/68	3210	3170
3	·11/68	2460	1
4	12/66	44	44
5	7/68	2390	269
6	8/67	1620	1622
7	1/69	4485	2540

The first and simplest way to look at these failures is to plot the number of failures against the date (Figure 8.5). This may be misleading, because all the components were not introduced at the same date, and so to understand the situation better the good ones (the unfailed components) must be looked at as well as the bad ones (the failed components). The data for the unfailed components is often not as readily obtainable as for the failed components. However, in this case it was possible to retrieve the *number of hours since new* and the *number of hours since last build* for the unfailed components which are analysed in Tables 8.3 and 8.4. The middle of Table 8.4 has been omitted so that the reader can check that he or she is able to carry out this calculation. Better still, if there is any data available for a product with which the reader is familiar, then now is the time to try establishing a failure pattern. It should be noted that the number of units exposed in terms of *hours since new* are less than those in terms of *hours since last build* because the same component can pass through each *hours since last build* lifeband several times over.

The results of this analysis can now be plotted (Figures 8.6 and 8.7) and there is a chance of establishing a more meaningful plot than the simple *No. of failures v date of failure*. The analysed data is better than Figure 8.5 because;

(1) it takes account of all the good (unfailed) components that are in service.
(2) it corrects for the fact that customers took delivery of their equipment at different times, and therefore some equipment has been running for longer than has other equipment.

Table 8.3

Customer	1-200	201-400	401-600	601-800	801-1000	1001-1200	1201-1400	1401-1600	1601-1800	1801-2000	2001-2200	2201-2400	2401-2600	2601-2800	2801-3000	3001-3200	3201-3400	3401-3600	3601-3800
	No. exposed, in 'Hours since last build'																		
1	10	9	12	19	18	21	20	21	21	8	15	35	17	6	1	2	0	0	0
2	2	0	0	3	0	0	1	1	1	1	1	1	1	0	0	2	1	1	0
3	2	6	1	3	3	4	0	3	1	2	0	1	1	1	1	0	0	0	0
4	2	3	0	2	3	0	1	2	1	4	0	0	0	0	0	0	0	0	0
TOTAL	16	18	13	27	24	25	22	27	24	15	16	37	19	7	2	4	1	1	0
Failures	2	1							1				2			1			
Cumulative No. exposed in lifeband	298	282	264	251	224	200	175	153	126	102	87	71	34	15	8	6	2	1	0
Proportion failed in each lifeband	2/298	1/282	0	0	0	0	0	0	1/126				2/34			1/6			
Proportion surviving in each lifeband	$\frac{296}{298}$	$\frac{281}{282}$							$\frac{125}{126}$				$\frac{32}{34}$			$\frac{5}{6}$			
Proportion of original sample Surviving	$\frac{296}{298} = R_1$	$\left(\frac{296}{298}\right)\left(\frac{281}{282}\right) = R_2$							$\left(R_2\right)\left(\frac{125}{126}\right)$ $R_3 = 0{\cdot}98189$				$1-\left(R_3\right)\left(\frac{32}{34}\right)$ $R_4 = 0.92412$			$\left(1-{}_4\right)\left(\frac{5}{6}\right)$ $R_5 = 0.77009$			
Cumulative Proportion of original sample	$1-R_1 = 0{\cdot}00671$	$1-R_2 = 0{\cdot}01025$							$1-R_3 = 0{\cdot}01811$				$1-R_4 = 0.07588$			$1-R_5 = 0.22991$			
Cumulative % of original sample failed.	0·7	1·0							1·8				7·6			23·0			

Table 8.4

Customer	No. exposed, in *'Hours since New'* 0-500	501-1000	1001-1500	1501-2000	2001-2500	2501-3000	3001-3500	3501-4000	4001-4500	4501-5000	5001-5500
1							4	16	24	25	11
2					1	3	1	5			
3				1	0	4	1	3	1		
4			1	0	3	0	0	3			
TOTAL			1	1	4	7	6	27	25	25	11
FAILURES	1			1	2		1		1	1	
Cumulative No. exposed in lifeband	114	113	113	112	110	104	97	90	63	37	11
Proportion failed in each lifeband											
Proportion surviving in each lifeband											
Proportion of original sample servicing											
Cumulative proportion of original sample failed	·009			·018	·037		·045		·060	·085	
Cumulative % original sample failed	0·9			1·8	3·7		4·5		6·0	8·5	

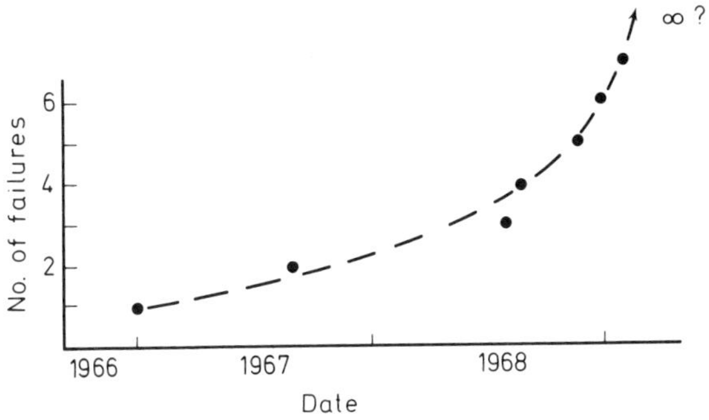

Figure 8.5 Raw failure data plotted against calendar date

The failure pattern must correlate with the engineering facts to be credible. The plot on a basis of *hours since last build* (Figure 8.6) does not look credible unless there are two failure modes present. The failed hardware when examined in the laboratory did not appear to have different failure modes to correlate with the two failure patterns. A theory explaining in terms of *time since last build* can be set aside

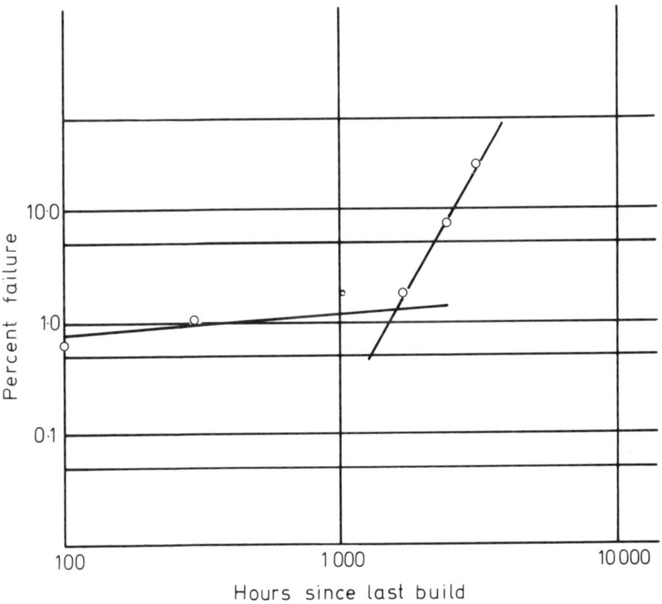

Figure 8.6 Cumulative percentage failed versus time since last build (Weibull scales)

as being inapplicable in the case of the example studied. The *hours since new* (Figure 8.7) look much more credible and suggests that the whole population is prone to failure.

The plots in Figures 8.6 and 8.7 are on Weibull paper[26] to which reference has already been made. More will be said in the next

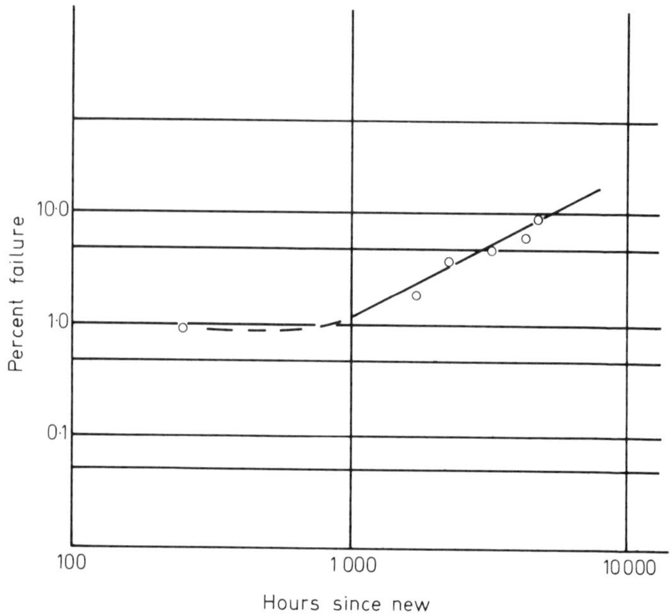

Figure 8.7 Cumulative percentage failed versus time since new (Weibull scales)

section. This is a very convenient set of scales to use since a wide variety of distributions give a reasonably linear plot. However, the correct method of analysis is more important than the scales chosen for plotting the results and if Weibull paper is not available, linear or normal/log probability papers can often be used successfully.

There are several other ways of handling data of this sort, for example, *ranking* which may be carried out as follows. The sample is first sorted into order. Each failure is assigned an order number in the following way. The increment is calculated from the formula:

$$\frac{N+1-(\text{previous order number})}{2+(\text{number of items after the current failure})}$$

where N is the sample size and an item is either a failure or a serviceable part. Then the order number of the failure is the previous order number (initially zero) plus the increment. For each failure the median rank (or cumulative probability of failure), is calculated from its order number by the formula:

$$\text{Median rank} = \frac{(\text{order number})-0{\cdot}3}{N+0{\cdot}4} \times 100\%$$

This is a good approximation; the exact calculations are rather complicated. The necessity for this juggling is that not all the units have run for the same amount of time, and therefore have not all equal opportunity to fail. But even this relatively straightforward calculation has an air of statistical sleight of hand which may not appeal to the reader. Where there is a reasonable amount of data available and no more than 20% of the population fails, it is probably best to use the *cumulative population surviving* approach shown in Tables 8.3 and 8.4. In this way the sceptics may be convinced because they can see how the answer is derived. This approach is more likely to appeal to engineers and managers.

8.5 WEIBULL DISTRIBUTION

Discussions at conferences often seem to suggest that the Weibull distribution provides the answer to every question on reliability. It does no such thing, but it is very convenient since many distributions that occur in practice can be simply approximated by the Weibull distribution.[26]

The Weibull equation is:

$$\begin{aligned} F(t) &= 1-R(t) \\ &= 1-\exp-\left[\frac{t-t_0}{\eta}\right]^{\beta} \end{aligned}$$

where $F(t)$ = cumulative probability of failure

$R(t)$ = cumulative probability of survival

β = Weibull slope parameter

η = characteristic life

t = random variable time (cycles, starts, number of machine tool passes etc.)

t_0 = origin of the distribution.

A blank chart with Weibull scales is shown in Figure 8.8 on which

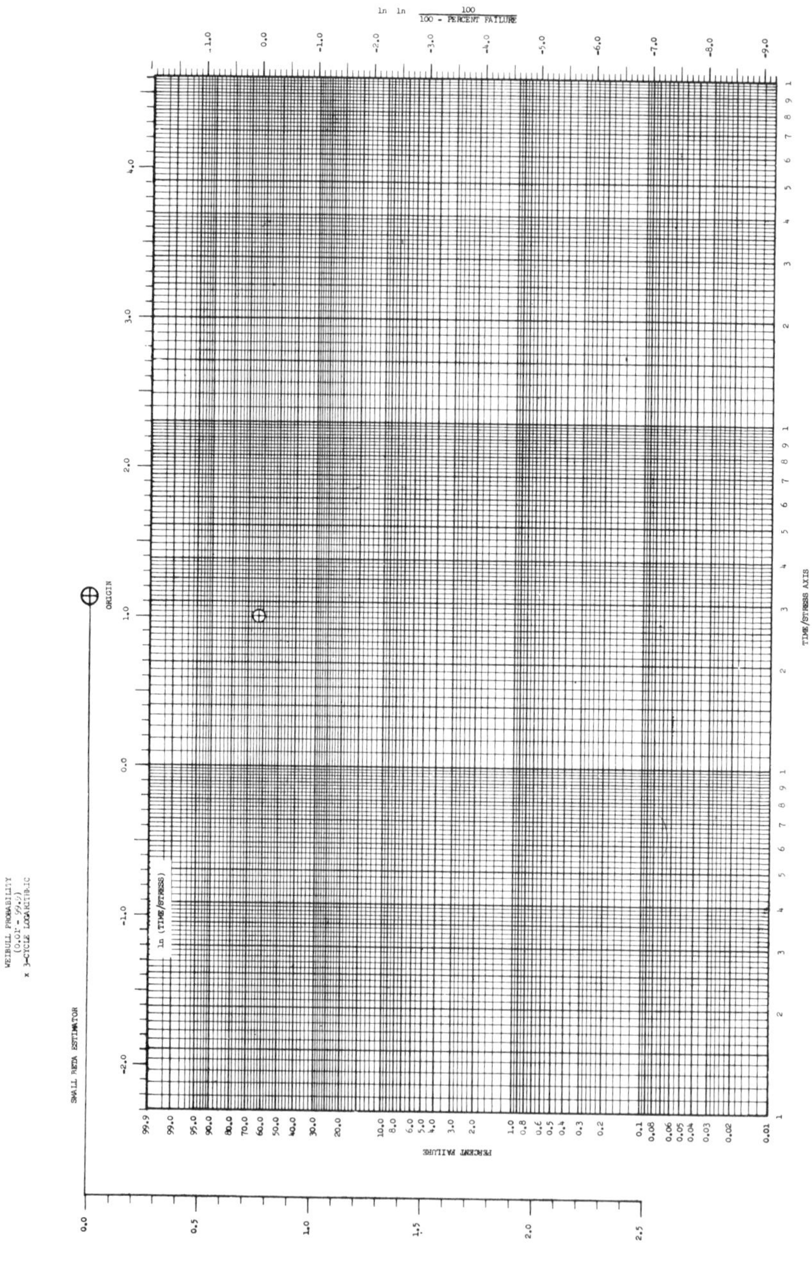

Figure 8.8 Weibull Probability Paper[27]

the Reader can experiment and discover the versatility of the function.

The probability density function $= f(t)$

$$= \frac{dF(t)}{dt}$$

$$= \frac{\beta(t-t_0)^{\beta-1}}{\eta^\beta} \exp - \left[\frac{t-t_0}{\eta}\right]^\beta$$

Plotting $f(t)$ against f for $\beta = \frac{1}{2}$, 1, $2\frac{1}{2}$, 3·44 as in Figure 8.9 shows that for different values of β the distribution can be approximated to exponential ($\beta = 1{\cdot}00$) or normal ($\beta = 3{\cdot}44$) as well as many other distribution curves. The value of β can be read off by eye if data is plotted on Weibull probability paper, and a fair approximation to

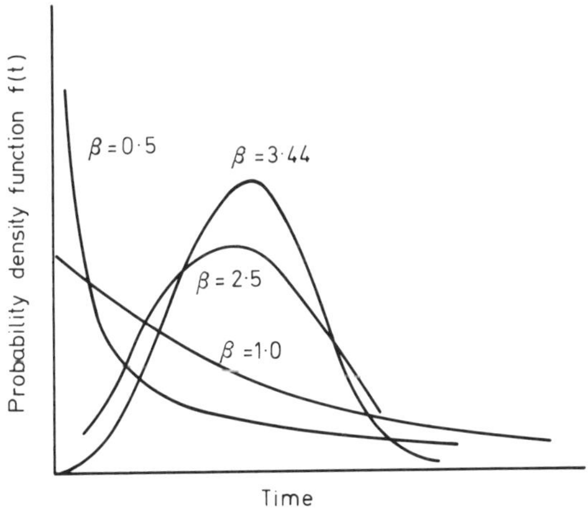

Figure 8.9 Failure distribution curves for different Weibull slopes β

the true distribution can be established rapidly. It can be seen if the failure rate is decreasing, constant, or increasing with time, which is usually the main fact required when failures are occurring in service. In the example of Figure 8.6 β is initially less than one (failure rate decreasing with time), and then changes to β much greater than one (failure rate increasing with time).

In the example of Figure 8.7, β is just greater than one, and so the failure rate is increasing slowly with time.

Because of the range of distributions covered by the Weibull model, it is usual for data to plot in a reasonable manner on Weibull probability paper. Extrapolation and interpolation is therefore possible in many cases without losing credibility and with an enhanced degree of confidence in the prediction.

If failures can occur at any time above $t = 0$ then the distribution will often plot as a straight line. If failures can not occur at $t = 0$, but only after some minimum time t_0, then t_0 can be established by trial and error as that value of t_0 which transforms the curve on Weibull paper into a straight line (Figure 8.10). This can be of real

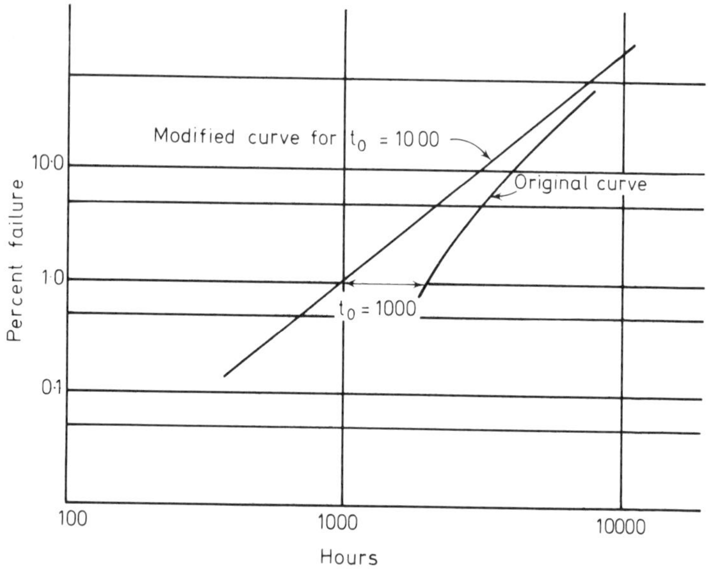

Figure 8.10 Determination of the initial safe life t_0 using Weibull probability paper

practical value, for if it is possible to establish a safe life, this allows the customer to operate his equipment up to time t_0 with complete confidence that no failures will occur.

η is the characteristic life, since for all values of β, 63% of the units will have failed when $\eta = t - t_0$ or when $\eta = t$ if $t_0 = 0$. This is not often useful other than to compare alternatives in a simple manner at the development stage, since most operators are not prepared to have 2/3 of their equipment fail.

Weibull's distribution is the greatest thing for those of us who do not enjoy statistics very much. Using it we can make simple statistical judgements with very little labour or effort.

8.6 SERVICE DATA IN PERSPECTIVE

Service data analysis must be seen in perspective. It is not a replacement or substitute for experienced trained service engineers, but is a management information system to ensure that the customer gets the product support he requires.

The analysis system must therefore be under the control of the Product Support Manager, or Service Manager and must be credible. It is no good setting up a system in the pious hope that it will somehow do some good. The system must be designed to provide just enough information analysed to enable the correct decision to be made. Unfortunately large expensive systems have come into being which provide employment for System Analysts and Programmers, but doing precious little to help the product support operation.

Decide what information you must have, and then move heaven and earth to get just that, and no more.

9

Modelling and simulation

9.1 SIMULATION

Simulation is a mathematical technique to investigate the effect of coupling together distributed phenomena. When the phenomena are not distributed, then the answer can be determined precisely. If a machine tool and operator combination always produce scrap for the first ten minutes on a Monday morning then the reliability of the machine and operator is completely determinate, being 0% for ten minutes and 100% thereafter. In the more common situation where the machine and operator combination produce, on average, 99 good components for every 100 components produced, the situation is less determinate. The unsatisfactory components in a batch of 1000 may be numbers 9, 84, 101, 102, 106, 299, 504, 605, 900, 977. This is an indeterminate situation, since it is not known when the bad components will appear. The combination of a number of indeterminate operations, some in series and some in parallel, that occurs in a typical production operation will be indeterminate. If the distribution of performance of the individual elements in the system can be determined, and these distributions sampled in the correct sequence to follow the sequence of events that occurs in the real situation, then the system is being simulated. The answers from this simulation will vary from one run to the next, but the distribution arising from a large number of simulation runs will represent the picture of events that can be expected in reality over a period of time. Simulation is a way of producing an answer when the situation is indeterminate. If 'average' reliabilities were used for each element of the system, the system would be determinate and an average answer would be produced. The determinate answer produced in this way would give no indication of the variability to be expected. The average reliability of an operator for which the daily reliability on three successive days is 94%, 95%, 96% is the same as for an operator with a daily reliability on three successive days of 90%, 100%, 95%.

It may be important to know what is the variability. Simulation can provide the answer.

9.2 WHEN AND BY WHOM SHOULD MODELS BE USED

The whole aim of modelling or simulating a system is to be 'wise after the event' before the event has even started. There are great claims made for the benefits of simulation in optimising systems. However, there is always a sneaking suspicion that much of the literature is written by people who have set up the models, and in particular have been associated with the mathematics and programming aspects, rather than by managers and executives who have used models to establish a particular way of operating. In other words there is a very real possibility that models can on occasions become an end in themselves rather than being a simple tool to achieve an end. A model should be regarded in the same way as a slide-rule—it is there to be picked up and used, like any other tool, to help reach a conclusion and provide an answer.

No amount of modelling can improve or degrade the inherent reliability of components or processes in a system. What a model can do is to help put these components into the correct sequence and help assign quantitative values to them. For example, if there are 6 lathes in 6 different machine shops the reliability of each lathe is unaffected by whether there is a spare tool in the cabinet beside each lathe or whether there are 6 spare tools in the central store. The apparent reliability of the manufacturing process of turning could be very different for the two cases considered if reliability is defined in its usual manner, that is, the probability of a piece of equipment being available for use in its designed manner during a given period in time. The availability of spare parts can influence the amount of time for which the equipment is available for immediate use. In simple cases there is clearly no advantage in considering a complicated model and a Manager's intuitive guess is probably the correct decision.

The system does not need to be complicated before it becomes difficult to assess mentally the effect of different arrangements. If a taxi fleet operator has ten taxis, it is not immediately obvious whether the reliability of the service to his customers will be improved or degraded by operating with one depot, two depots or even no depots at all but operating a radio controlled fleet. This in itself is a very simple example and in a more complicated example it would be impossible to quantify decisions and strategies without resorting to some form of model.

Models are for the cautious Manager because there is always a temptation to avoid making decisions based on intuition if a model appropriate to the system can be written. In this way it is possible to be cautious by being wise after the event before the event has started. Models provide the perfect excuse for avoiding any decision today and putting off difficult problems until tomorrow. Equally well models can be used to prove almost anything. This is not the way in which simulation and models should be used. If biased or incorrect assumptions are made for the various elements in the system, then the model will provide equally biased and incorrect answers. For a model to provide a meaningful answer it must be based on rational and realistic assumptions and on the maximum amount of factual data. In fact the biggest single advantage to come from establishing a model to simulate the apparent reliability of a system is that strenuous efforts must be made to establish the reliability of the different components of the system. In many cases the initial processing of this information may be sufficient to enable an optimum system to be established without the necessity for a complicated model. For example, in a transport fleet operation, by unearthing the sordid facts it may well be the case that unnecessarily long delays are occurring in returning components for repair and that by establishing an express parts delivery service the apparent reliability of the system can be improved to a level sufficient to meet all reasonable demands.

Collecting data is not an exercise for the faint hearted. So often the initial response to a request for detailed information will be that it is not available, or that it is too difficult to make it available. It will be necessary to ignore these remarks and attack the filing systems personally in order to show that the information is there for those prepared to dig deeply. It is true in my experience that many arguments about why the data cannot be provided last for longer than it would have taken to unearth the data. The data should be in the form of probability distributions wherever possible since this is the way events happen in practice. The information from a model will be less useful if average quantities, average times or average reliabilities are assumed for every component in the system.

9.3 DO YOU BELIEVE THE ANSWERS?

Before setting out on establishing a system model to simulate a service or production operation it is vital that the assumptions made must be agreed by or, better still, be established by the Manager who is going to make decisions based on the answers the model provides.

It is no use leaving the assumptions to be made to the computer programmer who will no doubt make the best assumptions possible but whose judgement can be questioned later. Their judgement will only be questioned of course when the simulation provides a disagreeable answer. Modelling is a complete waste of time and money if the assumptions are not agreed at the outset. If the answers are not the ones which the boss wants he will ignore them and justify this on the basis that he does not agree with the assumptions. And he would be right.

Equally well, the Manager who is going to use the model must also define the scope of the model with its boundaries and acceptable approximations. It is fascinating and technically delightful to sit in a corner and develop an all-singing and all-dancing model which takes account of every little element in the system. Whilst this is technically fascinating it is economically disastrous. It does happen.

This is enough general discussion on models and much more will be learned by spending an hour or so actually trying to write one. Large models are strictly for the experts using special modelling or simulation languages (e.g. SIMON) but the principles can be understood and the shortcomings realised by playing with a hand driven model for an hour or two. If you have not been involved in the modelling business and wish to develop some feel for its potential, an hour spent with the example in section 9.4 will improve your understanding of the simulation process.

The situations where modelling can be used effectively to improve the apparent reliability is in situations involving the number of spares held, the availability of spares, the availability of maintenance personnel, the location of personnel, the chances of a 'fix' being a cure, and the possibilities of changing elements in the system for ones with superior or inferior reliability. Different permutation of the system elements can be investigated before embarking on the expense of doing things in real life.

9.4 PUTTING THE PRINCIPLES OF MODELLING INTO PRACTICE

A fleet of vehicles carrying Service Engineers is required to operate in such a way that three quarters of the calls for assistance in a large city can be answered in less than one day. The reason for the requirement is that the competitors justifiably claim to be able to answer half of the calls for assistance the same day.

Here then is a firm operational reliability requirement which can be satisfied in many ways, some more expensive than others. Contributory factors to reliability level are:

(1) number of calls for help
(2) time to communicate with service engineer
(3) number of service engineers and vehicles
(4) disposition of the vehicles
(5) availability of the vehicles (vehicle reliability)
(6) availability of the service engineer (employee reliability)
(7) time to complete the service call
(8) availability of spare parts
(9) service calls where caller is out, hoaxes etc.
(10) revisit required (lack of employee skill)

The answers from a simulation can be no better than the assumptions made in setting up that simulation, the extreme case being GIGO (garbage in, garbage out).

For the answer to be meaningful the following information breakdown is required:

(1) no. of calls for help

This should be already available from the existing reports turned in by the service engineers and is shown in Figure 9.1. The line shown is

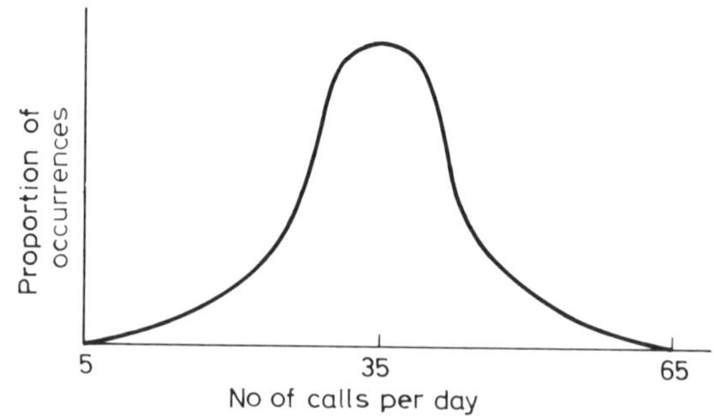

Figure 9.1 Distribution of the number of service calls made in a day

a normal distribution with a mean number of calls per day equal to 35, and a standard deviation of 10 (Figure 9.2).

This shows that on an average day there will be 35 calls, but that on 0·27% of occasions there will be less than 5 (or more than 65) calls.

(2) time to communicate with service engineer

This depends on the method of communication. Suppose that radio telephones are not fitted, and that the service engineer telephones

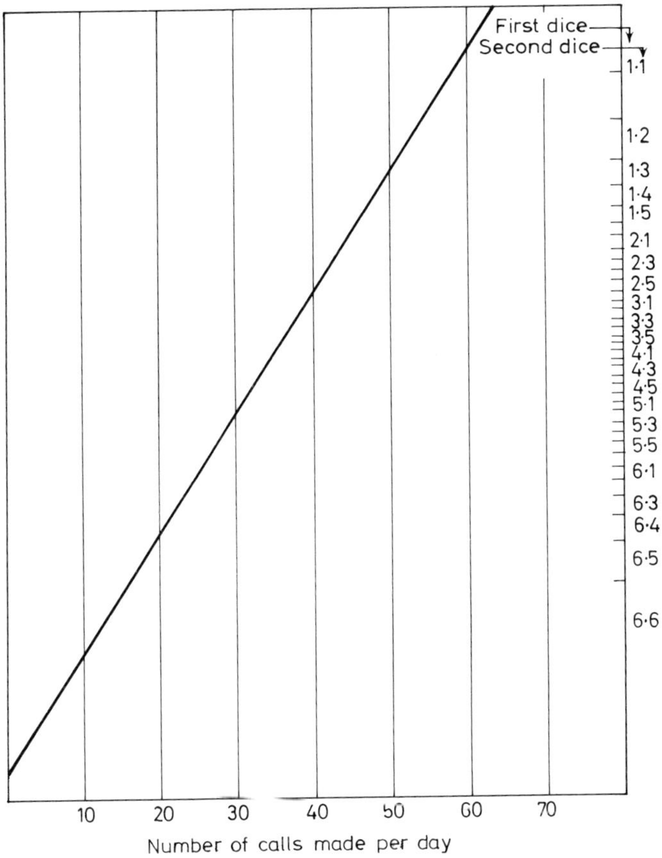

Figure 9.2 Distribution of the number of call made in a day presented for sampling by throwing two dice

in on completion of his job, or at 1/4 hourly intervals when not working. (Clearly alternative instructions could be tried in later simulations using 10 or 20 minute intervals for example).

(3) no. of service engineers and vehicles

Three of each at present; the decision whether to buy another vehicle or employ another engineer depends upon the outcome of the simulation.

(4) disposition of the vehicles

This information (for the existing instruction to drive to the nearest telephone kiosk and wait), is available from the service engineer's logbooks, see Figure 9.3. This is normal distribution again and is in

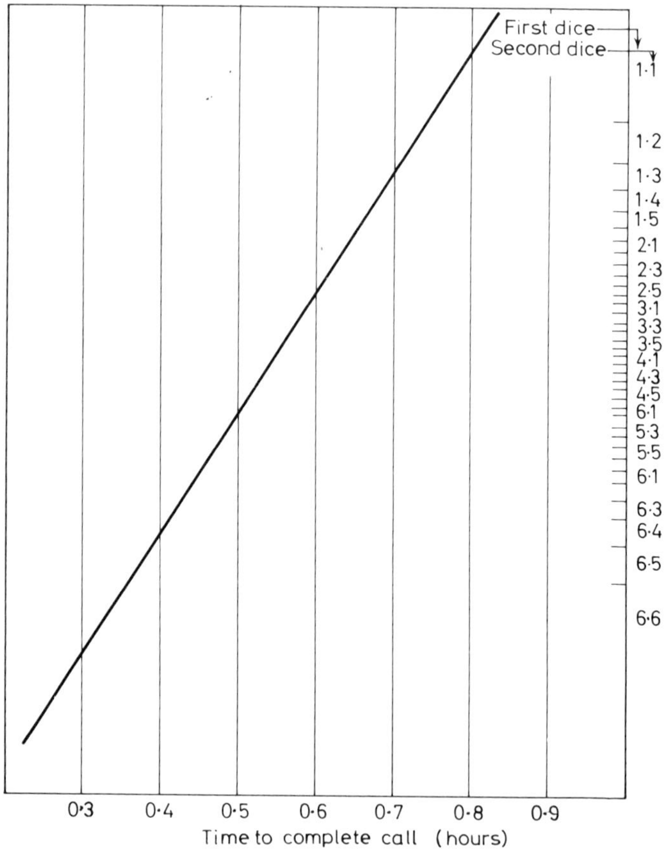

Figure 9.3 Distribution of the time taken to reach a customer presented for sampling by throwing two dice

terms of hours to reach the customer after leaving previous customer.

mean time from kiosk to customer = 0·55 hours
standard deviation = 0·1 hours

(5) availability of vehicles

Reliability of vehicles at the present time is that they are off the road for maintenance and repair 1 day in 50. The distribution of these days is not available, but with only 2% unreliability it will not influence the result of the simulation very much.

(6) availability of engineers

Including holidays, sickness, third party illness, the attendance records show that the engineers are available for work 19 days out of 20. Unfortunately the records do not show the distribution of time off, that is the number of consecutive days off.

(7) time to complete the service call

This information is available from the invoices which show the labour time and charge, see Figure 9.4. Yet again this is a normal distribution with a mean of 0·55 hours and a standard deviation of 0·1 hours. This includes the time to reach the telephone kiosk after completing the service call.

(8) availability of spare parts

The log books show that 1 in 6 times there is a delay (average 1 hour) whilst a spare part, not carried in the vans, is fetched from the depot.

(9) caller out, hoaxes etc.

Records show 1 call in 20 cannot be satisfied as unable to find address, caller gone out, hoaxes etc. and the average delay is 1/2 hour.

(10) revisits because fault not cured

Records show that 1 call in 20 requires a second visit to cure the fault (so the simulation should be re-run with a consecutive call in this case, but the second call should not count as another call).

(11) working day

Eight hours plus a maximum of 3 hours overtime. Meal breaks are taken during waiting time and so can be ignored.

Using the data now supplied, the reader can hand-drive the simulation to better understand how it all works, to realise the shortcomings of some of the assumptions, and to see the strange answers that come out after a short number of runs, before the simulation has settled down.

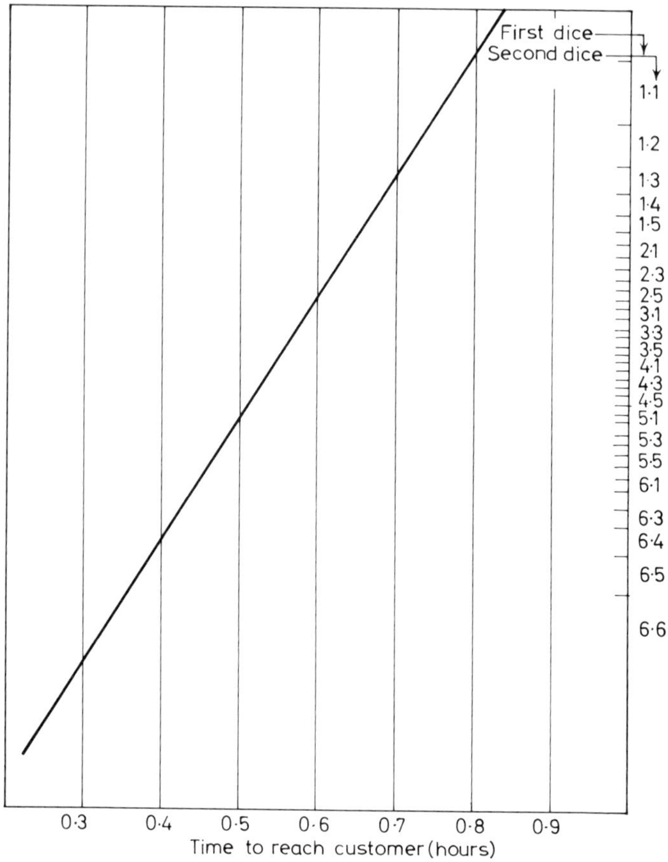

Figure 9.4 Distribution of the time to complete a call presented for sampling by throwing two dice

The simulation proceeds as follows:

(a) Establish the number of calls on day 1, throwing 2 dice and reading off the distribution on Figure 9.2.

(b) Establish no. of vans and engineers available on day 1 by marking 19 blue cards Y (= at work) and 1 blue card X (= absent) and drawing a card from the pack of 20 to establish if any of the 3 engineers is available. The drawn card must be returned, the deck shuffled, and re-drawn for each of the 3 engineers. A similar exercise using 50 white cards (49 marked Y and 1 marked X) decides how many of the 3 vans are available on day 1.

(c) Start by assigning a call to each available vehicle.

(d) Sample Figure 9.3 as in (a) to establish travelling time to the first caller and mark the time on the activity chart for that vehicle (Figure 9.5). Sample as in (b) on blue cards to establish if the caller is in. If out, got to (c) otherwise go to (e).
(e) Sample time to complete call (Figure 9.4) again by throwing 2 dice and then mark the time on activity chart (Figure 9.5) for that vehicle.

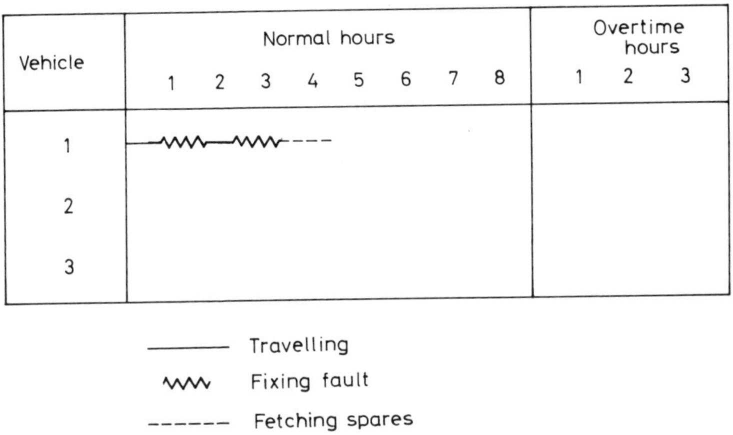

Figure 9.5 Method of plotting simulated activities of service engineer

(f) Sample availability of spare parts as (b), but using 6 yellow cards (5 marked Y and one marked X). Mark time to fetch spare part on vehicle activity chart (Figure 9.5) as appropriate.
(g) Check if fault cured, sampling as in (b) on blue cards. If not cured, go to (d) but do not count as another call. If cured go to (c) and start next call. The vehicles must be kept approximately in step so as to establish if overtime is required at the end of the day. After all time is expired, including overtime, establish proportion of calls satisfied in the day.

After a large number of simulation days it will be possible to reach a firm conclusion on the reliability of the operation (within the limits of the initial assumptions made).

Clearly in practice it is desirable to perform simulation on an analogue or digital computer in order that a large number of runs can be performed in a short time. The purpose of this hand exercise is to establish the principles and to show the possibilities as well as the limitations.

The results of about one hour's hand driving of this simulation are

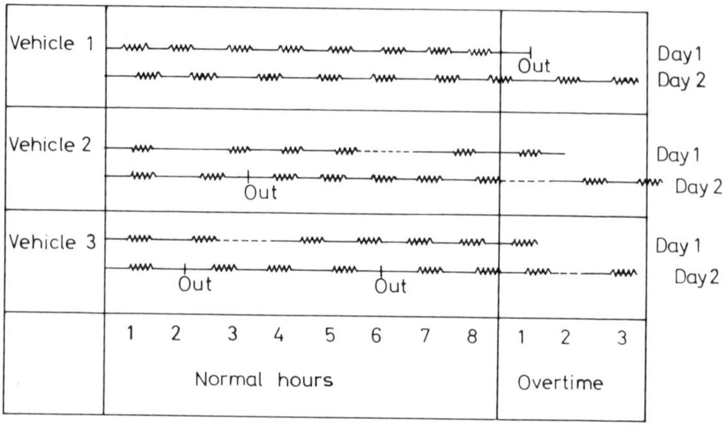

Day	Calls		Over-time hours
	Received	Answered	
1	22	22	2·6
2	41	29	9·7

Figure 9.6 Results of first two days simulated activities of the three service engineers and their vehicles

shown in Figure 9.6. It can now be seen that there is some hope of matching the performance in terms of the number of calls answered in one day.

The reader will have proved the great labour involved in a simple simulation such as this. A computer will remove the laboriousness of the simulation, but if the computing time is being bought, simulation can be expensive if the size is not restricted. Where the computer is owned by the simulator, and where running time is less important, the simulation can be more lavish and the time spent on producing an economical programme reduced.

9.5 WHAT OF THE FUTURE?

Less common in reliability circles is the war games aspect of modelling. Playing war games with the following combatants:

(1) operators *v* suppliers
(2) design organisations *v* manufacturing organisations
(3) inspection and quality *v* manufacture
(4) spares department *v* field service department

can be a fast and informal method of evaluating the affect of decisions. As an example, it may well be wrong to assume that the provision of a modification will appease the supplier and solve his problem. He may choose not to buy the modification at the offered price but to buy alternative equipment which he (possibly incorrectly) assumes is better. He may in the end accept what he has, which therefore means that the development money spent on producing the modification has been wasted. The possible series of events following an action or event can be explored by war game simulation using real people to take on the required roles.

Similarly a war game between design and manufacturing will help highlight the problems of both areas in attempting to improve the reliability of a piece of equipment. Unfortunately, some coordination meetings deteriorate into a form of war game and it seems a pity that these are not often used in a constructive fashion by acknowledging at the outset that a war game is about to start. The game can be pre-arranged to last for say one hour, rather than the more usual occurrence of holding a meeting which deteriorates into a form of war game without the combatants being able to remain dispassionate about the conclusions when the game is over. War games are quite common in the marketing field and it seems possible that there are real advantages to be gained by using this technique to thoroughly assess the results of adopting a particular strategy towards meeting the system reliability.

The other advantage of war game simulation is that different departments will learn how each other operates. The design department will be horrified to find out that the Buying department switch suppliers without telling anybody and without testing the reliability of the new product. The Production department will have their opportunity to slow down the Design department's enthusiasm for making better designs which disrupt production schedules. The Service department will have a chance to see whether their reports on in-service unreliability ever reach the action file in the Design department. The Designers will be made aware of the brutal ways in which their precious designs are being assembled on the shop floor.

Simulating is stimulating because it cannot be done unless you find out the facts. Unreliability is so often due to some breakdown in communcation rather than due to an insurmountable technical problem. A simulation exercise forces all parties concerned to face the facts.

Answers

CHAPTER 4

(1) $R_{total} = R_1R_2R_3R_4R_5R_6R_7$ (Figure 4.7).
(2) $R_{total} = 1-Q_{total} = 1-(1-R_1R_2)(1-R_3R_4)$ (Figure 4.8).
(3) $R_{total} = 1-(1-R_1R_2)^2$ (Figure 4.9).
(4) $R_{total} = 1-(1-R_1)^3$ (Figure 4.10).
(5) $R_{total} = R_1(1-(1-R_2)^2)R_3R_4(1-(1-R_5)^2)R_6R_7R_8R_9R_{10}$. (Figure 4.11).

CHAPTER 5

(1) If B no better than C, it appears to be better once every three tests.
(2) If B no better than C, it appears to be better (both Bs above both Cs) once every six tests.
(3) If B no better than C, it appears to be better (all Bs better than all Cs) once every ten tests.

CHAPTER 6

(1) 0·27; 0·004; Yes
(2) 0·03; a little
(3) 1; fairly
(5) 0·37

REFERENCES

1. *Genesis*, ch. 6, v. 14.
2. BROOK, R. H. W. and PARRY, J. S. C., 'Cumulative Damage in fatigue—a step towards its understanding', *J. Mech. Eng. Sci.* **11** No. 3, 1969.
3. FREUDENTHAL, A. M. and GUMBEL, E. G., 'Distribution functions for the prediction of Fatigue life and Fatigue strength,' *Int. Conf. on Fatigue of metals.* Inst. Mech. Engrs., London, 1956.
4. WEIBULL, W., 'A statistical distribution function of wide applicability', *J. appl Mech.*, **18** 293, 1951.
5. NIXON, F., 'Testing for satisfactory life', ASTM, STP 423, 1967.
6. CICCI, F., 'An investigation of the statistical distribution of constant amplitude fatigue endurance for a maraging steel', *Inst. of Aerospace Studies, Univ. of Toronto*, UTIAS Tech Note No. 73.
7. GOODMAN, J., *Mechanics applied to engineering*, Longmans, Green & Co., 1899.
8. PASSEY, R. D. C., 'The effect of mean stresses on fatigue strength', *Aircraft Engineering*, June 1969.

9. BROOK, R. H. W. and SNOW, E. W., 'The influence of indeterminate vibratory loads on designing for reliability at finite lives', *Aeronautical Research Council paper* No. 32–645, 5 Feb. 1971.
10. BOMPAS-SMITH, J. H., 'The determination of distributions that describe the failure of mechanical components', *Eighth conference on Reliability and Maintainability*, ASTM and ASME, July 1969.
11. NIXON, F., 'Quality and Reliability', *Engineering*, 11 April 1969.
12. BROOK, R. H. W., 'Reliability Prediction and Assessment', *The Quality Engineer*, **35** No. 1. 1971.
13. Pratt and Whitney Aircraft, 'Service Data Usage', ATA/IATA/AICMA Meeting Washington, U.S.A., May 1970.
14. SHAININ, D., *How to achieve productive experimentation*, Rath and Strong Inc., Boston, Mass., U.S.A.
15. FRENCH, M. J., 'Reliability tests with artificially damaged components', *The Engineer*, May 1965.
16. GOODMAN, R., *Teach Yourself Statistics*, English Univ. Press, 1957.
17. 'Information Exchange—Why and How', ATA/IATA/AIA/AICMA Data Exchange Steering committee, Washington, U.S.A. 1970.
18. SPRAKER, W. A., LOOMIS, J. P. and BAGBY, F. L., 'Trend analysis as applied to Gas-Turbine engines and aircraft mechanical equipment', *AIAA/SAE/ASME Aerospace Rel. and Maint. Conf.*, Washington, May 1963.
19. 'Mechanical Reliability Concepts', *ASME Design Engineering conference*, New York, 17 May 1965.
20. DAVIS, N. K., 'United Air Lines Test and Replace As Necessary (TARAN) Programme', *FAA Maintenance symposium*, Oklahoma City, 7 Nov. 1967.
21. HUDDIE, D. P., 'Economics of Propulsion systems for Air Transport', *Royal Aero. Soc. Lecture*, Dublin, 11 Feb. 1970.
22. WALLACH, R. A., 'The impact of Reliability and Maintainability on Airline Operation', *Los Angeles Maintainability Association*, 28 Nov. 1967.
23. ROBERTSON, K. D. and OLSON, B. D., 'The value of maintenance', *FAA Reliability and Maintainability symposium*, Oklahoma City, 8 Nov. 1967.
24. EDWARDS, L. S., 'Advances in Aviation Maintenance Technology', *FAA Reliability and Maintainability symposium*, Oklahoma City, 8 Nov. 1967.
25. BROOK, R. H. W., 'Planning for Failures', *I. Mech. Eng. conference on Safety and Failure of Components*, University of Sussex, Sept. 1969.
26. MITCHELL, R. A., 'Introduction to Weibull Analysis', *PWA* 3001, Pratt and Whitney Aircraft, East Hartford, Connecticut, U.S.A. 1967.
27. *Weibull Probability chart*, Chart No. 118 series, TEAM, 104, Belrose Ave., Lowell, Mass., U.S.A.
28. FISHER, R. A., *Statistical Methods for Research Workers*, Oliver and Boyd.

Index

Aero engine industry, 33
Aircraft, 32, 34–35
  delay cost, 35
Airframe structures, 23
Airlines, maintenance, 95–97
Alert levels, 100
Artificially damaged components, testing with, 73–75
Audit teams, 99

Binomial distribution, 80
  confidence limits for, 81–82
Binomial law of reliability, 51
Bolt failure problem, 7–8
Bridges, 23

Calculated risk, 7
Central heating system, 36
Checklists, 39–41
Coin tossing, 10, 76, 80
Computers, 96, 124
Confidence limits, assessment of, 80
  for Binomial distribution, 81–82
  for Poisson distribution, 82–85
  mathematical approach to, 79–86
Control system, failure analysis, 47–50
Corrosion tests, 72
Cost(s), development, 58
  maintenance labour, 92
  of ownership, 35, 36
  of unreliability, 35
Costing, whole life, 35
Cumulative population surviving, 109
Customer information retrieval, 98

Data analysis, collection of data, 116
Data analysis, *continued*
  fatigue failure problem, 104
  see also Service data
Decision making, 95, 103
  before test, 68
  statistical aids to, 79–80
  test result influence on, 73
Defect modes, 43–44
Design Co-ordination Manual, 37
Design reliability, 31–42
  changes in design, 34
    quantitative evidence for, 41–42
  specification, 31
  'swallow your pride and start walking approach', 38
Design review, 38
Designer, information available to, 37
Development, 58
  meaning of, 58–61
Development costs, 58
Development Engineer, 73, 74
Development problems and service problems, 69–71
Development testing, 58
  purpose of, 59
Differences in reliability, 71
Distribution, Binomial, 80
    confidence limits for, 81–82
  exponential, 52
  of diameters of component, 13
  of fatigue properties, 11, 15, 16, 24
  of hardness, 11
  of possible applied loads, 25
  of Ultimate Tensile Strength, 11, 15
  Poisson, 80
    confidence limits for, 82–85
  Weibull, 109
Distribution phenomena, 10–14
  and material properties, 15–30

Distribution phenomena, *continued*
  effect of coupling, 114
Distribution table, 14

Environmental conditions, 25, 26
Exponential distribution, 52

Factorial testing, 63–67, 72
Failure, conditions for occurrence or non-occurrence of, 74
  failure, 104
  indicator parameters, 88
  probability of, 56–57
  service, 74
Failure analysis, 43–50
  aim of, 43
  case for, 43–47
  control system, 47–50
  form headings for, 43
  person best suited for, 46–47
  quantitative assessments and estimates, 45
  results of, 46
  types of, 43
Failure mode, and effects analysis, 43, 49
  in overstress tests, 60
  on checklist, 39–40
Failure patterns, 96, 103, 104, 107
Failure rate, 5–7
  and age correlation, 96
  and maintenance interval, 90, 94
  and time since new, 95
Fallacy of the type test, 59
Fatigue, major and minor cycle effects, 20–21
  statistical nature of, 15
Fatigue failure problem, 104
Fatigue life scatter, 19
Fatigue limit, 19, 28
Fatigue properties, 28
  distribution of, 11, 15, 16, 24
  of alternative designs, 29
  scatter in, 15, 28
Fatigue stress/fatigue strength ratio, 74–75
Fatigue testing, artificially damaged components, 74
  overstress, 59–60
Finger troubles, 88
Finishing treatments, 72
Flexible shaft problem, 7

Gas turbine compressor aerofoils, 74
Gas turbine engines, 34, 102
  components, 22
  development problems, 70–71
Goodman diagram, modified, 19

Hardness, distribution of, 11
House building, 2, 33

KISS system, 31

Load distribution, 25
Load variability, 24–25

Maintenance, airlines, 95–97
  and reliability, 87–97
    qualitative aspects, 89
    quantitative aspects, 90
    theoretical relationships, 93–95
  inspection without disassembly, 96
  labour costs, 92
  on condition, 88
  scheduled, 96
Maintenance interval, and failure rate, 90, 94
  and reliability, 92, 95
Material, choice of, 28
Material properties, 11
  and distributed phenomena, 15–30
    effect on reliability, 19–24
  availability of data on, 26–28
  importance of, 29–30
Mathematical models, 50–57
Modelling, 114–125
  aim of, 115
  applications, 117
  assumptions made, 116–117
  defining scope of, 117
  principles and practice, 117–124
  service engineer call availability, 117–124
  war games, 124–125
Modification standards, 102
Motor car, 4, 5, 12, 32
  components, 23
  maintenance and reliability, 87

Noah's ark, 4
Non-destructive testing, 24

Numbers game, 6
  data required for, 8–9
  when to use, 6–10

On condition maintenance, 88
Overload cyclic tests, 29
Overstress factors, 60
Overstress tests, 59

Parallel system, 50
Passey range-mean diagram, 20
Poisson distribution, 80
  confidence limits for, 82–85
Poisson probability chart, 14, 82–85
Pratt and Whitney, 37
Pre-planning, 61, 68, 69
Probability of failure, 56–57
Probability chart, 82–85
Probability concepts, 6
Probability density function, 111
Probability distributions, 116
Probability paper, 108–112
Product Support Manager, 113
Production Engineer, 39
Production techniques, 26, 27
Project Definition Phase, 33

Quality and reliability, 1–4
Quality control, 24

Ranking, 108
Reliability, and maintenance, 87–97
    qualitative aspects, 89
    quantitative aspects, 90
    theoretical relationships, 93–95
  and maintenance interval, 92, 95
  and quality, 1–4
  apparent and theoretically obtainable levels, 102
  binomial law of, 51
  definitions, 4, 6
  differences in, 71
  in-service assessment, 101–103
  of components in parallel, 50
  of components in series, 50
  of process or design, 10
  overall assessment of, 102
  quantitative specification, 32
Reliability analysis, 43
Reliability Engineer, 39
Reliability logic, 50–57
  with no data, 55
Reliability tests, effects of permissible tolerance, 68
  transducer, 60
Reserve factors, 17

Safety factors, 17, 26
Series-parallel system, 55
  with identical paths, 55
Series system, 50
  with some stand-by parallel components, 55
Service data, aim of detailed analysis, 103–109
  analysis of, 98–113
    for problem areas, 100
  audit team, 99
  in perspective, 113
  reporting by service engineer, 99
  sources of, 98
  spares orders, 99
Service Engineer, 39, 86, 99
  modelling call availability, 117–124
Service failures, 74
Service Manager, 100, 113
Service problems and development problems, 69–71
Signal generators, wiring configuration assessment, 55–57
Simulation, 114–125
  aim of, 115
  assumptions made, 116–117
  definition, 114
  war games, 124–125
Spares orders, 99
Specifications, 26
  for reliable design, 31–33
  requirements, 32
Statistics in decision making, 79–80
  surface finish effects, 74

TARAN programme, 88, 95
Test programmes, designing, 61–67
  factorial testing, 63–67, 72
  importance of planning, 77
  looking for differences, 71
  pre-planning, 61, 68, 69
  reducing number of tests, 64
Test results, chances of correctness of, 76
  influence on decision, 73

Test results, *continued*
mathematical assessment of, 79
Test rigs, development, 58
Testing, with artificially damaged components, 73–75
see also under specific type of test
Time between overhauls, 42
Tolerances, permissible, 68
Tossing a coin, 10, 76, 80
Transducer, reliability tests, 60
Trend analysis, 100
Triplicated system, 55
Type test, fallacy of, 59

Ultimate Tensile Strength, distribution of, 11, 15
scatter plot of, 17
Unreliability, cost of, 35

War games, 124–125
Weibull distribution, 109
Weibull equation, 109
Weibull probability paper, 108–112
Weibull scales, 108
Whole life costing, 35